Kapil Gupta, Pallab Sarmah, Promod Kumar Patowari (Eds.)
Hybrid Manufacturing Processes

Also of Interest

Additive Manufacturing.
Science and Technology
Emrah Celik, 2025
ISBN 978-1-5015-2023-5, e-ISBN (PDF) 978-1-5015-2024-2

Hybrid Composite Materials and Manufacturing.
Fibers, Nano-Fillers and Integrated Additive Processes
Edited by: Vipin Kumar, 2025
ISBN 978-3-11-101934-5, e-ISBN (PDF) 978-3-11-101954-3

Smart Alloys.
In Energy, Biomedical, Engineering, Aerospace and Automotive Industries
Edited by: D. Lakshmi and A. Vallimanalan, 2025
ISBN 978-3-11-163005-2, e-ISBN (PDF) 978-3-11-163020-5

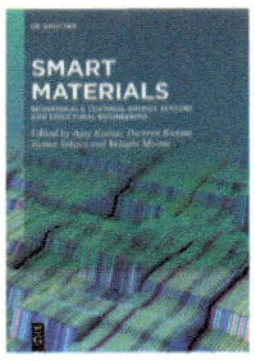

Smart Materials.
Biomaterials, Coatings, Energy, Sensors and Structural Engineering
Edited by: Ajay Kumar, Parveen Kumar, Tamer Sebaey and Velaphi Msomi, 2026
ISBN 978-3-11-151283-9, e-ISBN (PDF) 978-3-11-151571-7

Engineering Risk Management.
Optimizing Operational Risk Decision-Making, Safety and Reliability, Risk Assessment
Thierry Meyer and Genserik *Reniers,* 2025
ISBN 978-3-11-149058-8, e-ISBN (PDF) 978-3-11-149363-3

Hybrid Manufacturing Processes

From Micromachining to Additive Manufacturing

Edited by
Kapil Gupta, Pallab Sarmah and Promod Kumar Patowari

DE GRUYTER

Editors
Prof. Kapil Gupta
University of Johannesburg
Doornfontein Campus
Beit Street
Johannesburg 2028
South Africa
kgupta@uj.ac.za

Dr. Pallab Sarmah
University of Johannesburg
Doornfontein Campus
Beit Street
Johannesburg 2028
South Africa
psarmah@uj.ac.za

Prof. Promod Kumar Patowari
National Institute of Technology Silchar
Silchar 788010
Assam, India
ppatowari@mech.nits.ac.in

ISBN 978-3-11-170400-5
ISBN 978-3-11-170445-6 (PDF)
ISBN 978-3-11-170446-3 (EPUB)
DOI https://doi.org/10.1515/9783111704456

Library of Congress Control Number: 2026932140

Bibliographic information published by the Deutsche Nationalbibliothek
The Deutsche Nationalbibliothek lists this publication in the Deutsche Nationalbibliografie; detailed bibliographic data are available on the Internet at http://dnb.dnb.de.

De Gruyter and Walter de Gruyter GmbH are part of De Gruyter Brill.
www.degruyterbrill.com

Questions about General Product Safety Regulation:
productsafety@degruyterbrill.com

Cover illustration: kynny/iStock/Getty Images Plus
Printing: CPI books GmbH, Leck

MIX
Papier | Fördert gute Waldnutzung
FSC® C083411
www.fsc.org

Preface

In recent years, hybrid manufacturing has proven effective in solving complicated issues pertaining to process difficult-to-machine and functionally essential materials and manufacturing of sophisticated products. Hybridization of various manufacturing technologies is essentially required to fulfill the novel requirements related to product variety and quality and overcome the challenges pertaining to the cost and sustainability.

This volume aims to provide a complete overview of recent developments in novel and unique combinations of technologies and their applications in the rapidly increasing field of hybrid manufacturing. The book compiles cutting-edge research contributions that emphasize the integration of subtractive, additive, forming, casting, and modern machining processes, demonstrating how these synergies can lead to superior performance, improved surface integrity, and increased process sustainability.

Chapter 1 provides an overview of hybrid and metamorphic systems and how they match with sustainable manufacturing principles. Chapter 2 examines the machinability of modern dental ceramics using electrochemical discharge machining technology. Chapter 3 focuses on integrating electric discharge machining with auxiliary techniques to improve process efficiency, reduce tool wear, and improve the surface quality of difficult-to-machine materials. Advancements in hybrid finishing for optical applications are discussed in Chapter 4. Chapter 5 covers details on casting-based hybrid manufacturing processes, whereas forming-based hybrid manufacturing processes are discussed in Chapter 6. Chapter 7 sheds light on effective hybridization of 3D printing and precision machining techniques. For the post-processing of parts and components fabricated by additive manufacturing, various technological combinations are discussed in Chapter 8. Chapter 9 dedicatedly highlights the assessment of sustainability of hybrid manufacturing with the help of case studies. Finally, the book ends with Chapter 10, which presents some most recent developments in electrochemical hybrid machining processes for process capability enhancement and product quality improvement.

Collectively, the 10 chapters in this book demonstrate the wide range and depth of research now being conducted on the topic of hybrid manufacturing. They demonstrate the shift from traditional subtractive machining to a broader landscape that includes micromachining, precision finishing, and additive manufacturing. The book also emphasizes sustainability, process hybridization, and interdisciplinary integration as three critical pillars influencing the future of modern manufacturing processes. We sincerely hope this book will serve as a reliable resource for expanding the boundaries of hybrid manufacturing technologies and stimulating novel fields of research in the domain.

Kapil Gupta
Pallab Sarmah
Promod Kumar Patowari

 | https://doi.org/10.1515/9783111704456-202

Contents

List of Authors

Desmond Klenam
Next Frontier in Advanced Materials Laboratory
School of Chemical and Metallurgical Engineering
University of the Witwatersrand
Johannesburg, South Africa
desmond.klenam@wits.ac.za
Chapter 1

Rakesh Kumar
Department of Mechanical Engineering
National Institute of Technology Kurukshetra
Kurukshetra
Haryana, India
rakeshk.nitkkr@gmail.com
Chapter 2

Ravi Pratap Singh
Department of Mechanical Engineering
National Institute of Technology Kurukshetra
Kurukshetra
Haryana, India
singhrp@nitkkr.ac.in
Chapter 2

Sandeep Singhal
Department of Mechanical Engineering
National Institute of Technology Kurukshetra
Kurukshetra
Haryana, India
ssinghal@nitkkr.ac.in
Chapter 2

Pradeep Kumar
Department of Mechanical and Industrial Engineering
Indian Institute of Technology Roorkee
Roorkee
Uttarakhand, India
kumarfme@gmail.com
Chapter 2

Mustafizur Rahman
National University of Singapore
Singapore
mustafizur.rahman23@gmail.com
Chapter 2

Yash Yalamalli
Department of Industrial Engineering and Management
M S Ramaiah Institute of Technology
Bengaluru 560054
Karnataka, India
yashyalamalli@gmail.com
Chapter 3

Vibha Sastry H. K.
Department of Industrial Engineering and Management
M S Ramaiah Institute of Technology
Bengaluru 560054
Karnataka, India
vibhasastry1705@gmail.com
Chapter 3

Priyadarshan V.
Department of Industrial Engineering and Management
M S Ramaiah Institute of Technology
Bengaluru 560054
Karnataka, India
priyadarshan7@gmail.com
Chapter 3

Anshul Agrawal
Department of Industrial Engineering and Management
M S Ramaiah Institute of Technology
Bengaluru 560054
Karnataka, India
ashagr300@gmail.com
Chapter 3

Siddhartha Kar
Department of Industrial Engineering and Management
M S Ramaiah Institute of Technology
Bengaluru 560054
Karnataka, India
siddkar.nita@gmail.com
Chapter 3

 | https://doi.org/10.1515/9783111704456-204

Mahesh Kumar Sah
Materials Processing Research Laboratory
Department of Mechanical Engineering
Dr B.R. Ambedkar National Institute of Technology
Jalandhar 144008
Punjab, India
maheshk.me.24@nitj.ac.in
Chapter 4

Harpreet Singh
Materials Processing Research Laboratory
Department of Mechanical Engineering
Dr B.R. Ambedkar National Institute of Technology
Jalandhar 144008
Punjab, India
signhh@nitj.ac.in
Chapter 4

Orhan Çakır
Department of Mechanical Engineering
Yildiz Technical University
Istanbul, Turkey
orhanc@yildiz.edu.tr
Chapter 5

Çağla Özbilek
Department of Mechanical Engineering
Yildiz Technical University
Istanbul, Turkey
cagla.sen@std.yildiz.edu.tr
Chapter 5

Beyzanur Yanık
Department of Mechanical Engineering
Yildiz Technical University
Istanbul, Turkey
beyzanur.yanik@std.yildiz.edu.tr
Chapter 5

Miraç Mutlu
Department of Mechanical Engineering
Yildiz Technical University
Istanbul, Turkey
mirac.mutlu@std.yildiz.edu.tr
Chapter 5

Salih Yeşilsancak
Department of Mechanical Engineering
Yildiz Technical University
Istanbul, Turkey
salih.yesilsancak@std.yildiz.edu.tr
Chapter 5

Mohammad Zaher Akkad
Department of Mechanical Engineering
Yildiz Technical University
Istanbul, Turkey
mohammad.akkad@std.yildiz.edu.tr
Chapter 5

Orhan Çakır
Department of Mechanical Engineering
Yildiz Technical University
Istanbul, Turkey
orhanc@yildiz.edu.tr
Chapter 6

Çağla Özbilek
Department of Mechanical Engineering
Yildiz Technical University
Istanbul, Turkey
cagla.sen@std.yildiz.edu.tr
Chapter 6

Beyzanur Yanık
Department of Mechanical Engineering
Yildiz Technical University
Istanbul, Turkey
beyzanur.yanik@std.yildiz.edu.tr
Chapter 6

Miraç Mutlu
Department of Mechanical Engineering
Yildiz Technical University
Istanbul, Turkey
mirac.mutlu@std.yildiz.edu.tr
Chapter 6

Salih Yeşilsancak
Department of Mechanical Engineering
Yildiz Technical University
Istanbul, Turkey
salih.yesilsancak@std.yildiz.edu.tr
Chapter 6

Mohammad Zaher Akkad
Department of Mechanical Engineering
Yildiz Technical University
Istanbul, Turkey
mohammad.akkad@std.yildiz.edu.tr
Chapter 6

Vivek Kumar Tiwary
Department of Mechanical Engineering
KLS Gogte Institute of Technology
Udyambag, Belagavi
Karnataka, India
vgtiwary@git.edu
Chapter 7

Arunkumar P.
Department of Mechanical Engineering
KLS Gogte Institute of Technology
Udyambag, Belagavi
Karnataka, India
akp@git.edu
Chapter 7

Vinayak R. Malik
Department of Mechanical Engineering
KLS Gogte Institute of Technology
Udyambag, Belagavi
Karnataka, India
vrm@git.edu
Chapter 7

Abdul Wahab Hashmi
School of Mechanical Engineering
Shandong University of Technology
266 Xincun W Rd
Zhangdian District, Zibo
Shandong 255000, PR China
hashmicad@gmail.com
Chapter 8

Yebing Tian
School of Mechanical Engineering
Shandong University of Technology
266 Xincun W Rd
Zhangdian District, Zibo
Shandong 255000, PR China
and Shandong Key Laboratory of Precision
Manufacturing and Special Processing
Zhangdian District, Zibo
Shandong 255000, PR China
tianyb@sdut.edu.cn
Chapter 8

Mamilla Ravi Sankar
Department of Mechanical Engineering
Indian Institute of Technology Tirupati
Tirupati 517619
Andhra Pradesh, India
evmrs@iittp.ac.in
Chapter 8

Binnur Sagbas
Mechanical Engineering Department
Yildiz Technical University
34349 Besiktas
Istanbul, Turkey
bsagbas@gmail.com
Chapter 8

Mohit Kumar
Department of Mechanical Engineering
Indian Institute of Technology Bombay
Mumbai, Maharashtra, India
mohit_k@me.iitr.ac.in
Chapter 9

Tanmay Tiwari
Center for Precision Manufacturing
Department of Mechanical Engineering
University of Akron
Akron, OH, USA
ttiwari@uakron.edu
Chapter 9

Chandra Shekhar Rakurty
Research and Development
The M. K. Morse Company
Canton, OH 44707, USA
rakurtys@mkmorse.com
Chapter 9

Kapil Gupta
Department of Mechanical and Industrial
Engineering Technology
University of Johannesburg
Johannesburg 2028, South Africa
kgupta@uj.ac.za
Chapter 10

Desmond Klenam

1 Sustainable Hybrid and Metamorphic Manufacturing Systems: Trends and Perspectives

Abstract: The new era of sustainable manufacturing is shaped by digital innovation. This chapter provides a critical examination of hybrid manufacturing (HM) and metamorphic manufacturing (MM) technologies as transformative enablers of next-generation production paradigms, focusing on operational efficiency, process adaptability, and environmental sustainability. The HM approach integrates additive and subtractive processes within a unified system to achieve high-precision output, minimize material waste, and enable the fabrication of complex geometries seamlessly. Furthermore, MM leverages intelligent, real-time thermo-mechanical deformation guided by artificial intelligence and sensor-driven feedback to simultaneously design component geometry and tailor material properties. These technological trajectories signify a fundamental shift from conventional computer numerical control-based frameworks to highly responsive, cyber-physical production systems consistent with Industry 4.0 and 5.0 principles. Through a comparative analysis of successive digital manufacturing waves, this chapter situates these advancements within broad efforts toward circular economy practices, production resilience, and mass customization. Sector-specific case studies from aerospace, healthcare, and automotive domains underscore the practical relevance. The discussion concludes with strategic guidelines for implementation, addressing persistent challenges such as integration complexity, workforce capability gaps, and the need for standardized protocols.

Keywords: sustainable manufacturing, digital manufacturing, hybrid manufacturing, Industry 4.0, metamorphic manufacturing

1.1 Introduction

Sustainable manufacturing represents a critical shift in industrial production, merging environmental responsibility with economic viability. It addresses climate change, pollution, and resource depletion by reducing energy consumption, waste generation, and material use across the production life cycle [1]. Once a compliance concern, it is now a strategic imperative, offering cost savings, efficiency gains, and alignment with evolving regulatory and consumer priorities. It also advances global agendas like the UN SDGs and Africa Agenda 2063 [2], encouraging cross-sector collaboration, as illustrated in Fig. 1.1.

 | https://doi.org/10.1515/9783111704456-001

Manufacturing requires urgent transformation and modernization. It accounts for nearly one-third of global energy use, ~40% of greenhouse gas emissions, and generates over two billion tons of waste [1, 3–5]. In response to regulatory shifts and rising consumer demands, the sector must move from linear "take-make-waste" models toward circular, sustainable practices [6–8]. This shift is both ethical and vital for long-term competitiveness and sustainability [9, 10]. As manufacturing evolves beyond lean and Industry 4.0 automation, growing demands for flexibility, complexity, and sustainability are accelerating the adoption of next-generation paradigms, particularly hybrid manufacturing (HM) and metamorphic manufacturing (MM) [11–15].

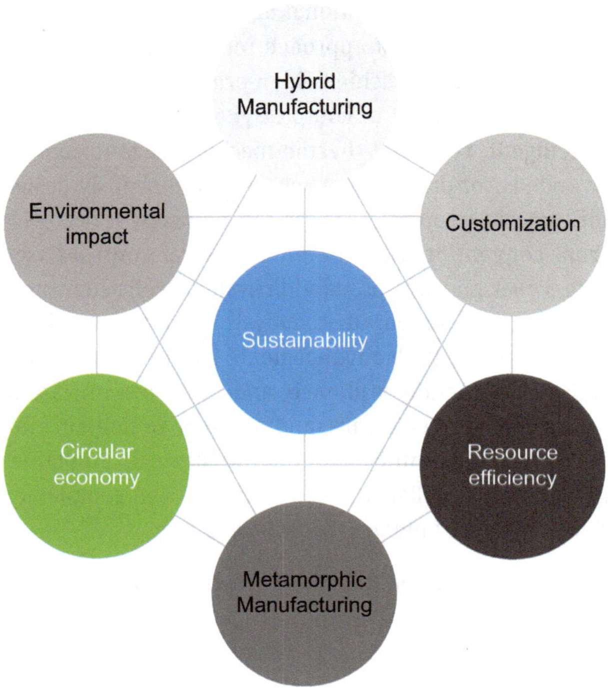

Fig. 1.1: The interconnectedness of the various aspects of the manufacturing spectrum in the digital age.

HM and metamorphic manufacturing systems (MMSs) represent advances in digital manufacturing evolution [11, 12, 16–18]. Generally, HM synchronizes additive and subtractive processes using digital tools for precision, efficiency, and real-time adaptability [19–25]. Additive techniques reduce material waste [10, 23, 26–31], while subtractive methods ensure high-quality finishes [32]. This convergence enables complex designs, energy optimization, and defect minimization, positioning HM as a critical bridge between the second and third digital manufacturing waves. The concepts that distinguish HM from MMS are summarized in Tab. 1.1. Building on HM, MMS integrates artificial intelligence (AI) and cyber-physical systems (CPS) to enable self-reconfiguring, intelligent factories [11, 12]. Often referred to as robotic blacksmithing, MMS emphasizes modularity [33], reconfigurability [34], and embedded intelligence [11, 35], adapting autonomously to dynamic production needs. Both paradigms advance the capabilities of computer numerical control (CNC) and additive manufacturing (AM) [10, 23, 26, 36–39], supporting sustainable, resilient, and customizable production systems [35, 40–47].

Tab. 1.1: Distinguishing features between HM and MM.

Concept	Hybrid manufacturing	Metamorphic manufacturing
Core role in digital manufacturing	Extends digital manufacturing by integrating digital design, monitoring, and control across multiple processes in one system	Represents the next leap: full cyber-physical adaptability and intelligent system reconfiguration
Digital backbone	Relies on CAD/CAM, sensors, and real-time control systems to coordinate additive/subtractive actions	Depends on AI, digital twins, modular design, and machine learning for self-driven system evolution
Level of digitization	Digital control of discrete integrated processes	Full digitization of system architecture, logic, and physical form
Connection to Industry 4.0	Embodies early-to-mid Industry 4.0 by integrating smart manufacturing techniques	Aligns with Industry 5.0 ideals: human-machine collaboration, adaptability, and resilience
Feedback and intelligence	Uses digital feedback for process optimization	Uses closed-loop intelligence to adapt both process and structure autonomously

Integration of Industry 4.0 technologies such as IoT, AI, and digital twins further enhances HM and MMS by enabling predictive maintenance, real-time optimization, and simulation-driven design [1, 10–12, 26, 36, 37, 39, 48–50]. These tools improve transparency, resource management, and supply chain collaboration, ensuring intelligent compliance with evolving environmental standards [15, 39, 51].

This chapter critically explores how hybrid manufacturing and MMS contribute to sustainable production. It defines core concepts, assesses their capabilities and limitations, and examines strategies for integrating sustainability. The discussion further reviews technological trends, outlines future directions, and identifies research gaps, offering guidance for advancing resilient, eco-conscious manufacturing.

1.2 Foundational Concepts in Advanced Sustainable Digital Manufacturing

Advanced sustainable manufacturing integrates digital technologies, circular economy principles, and systems thinking. It marks a shift from CNC (first wave) to hybrid (second) and metamorphic (third) manufacturing paradigms (Fig. 1.2). Core enablers include computer-aided design (CAD), digital twins, and AI-powered CPS, which support real-time monitoring, optimization, and predictive maintenance. These tools drive efficiency, minimize waste, and enable data-driven, adaptive decision-making across the product life cycle.

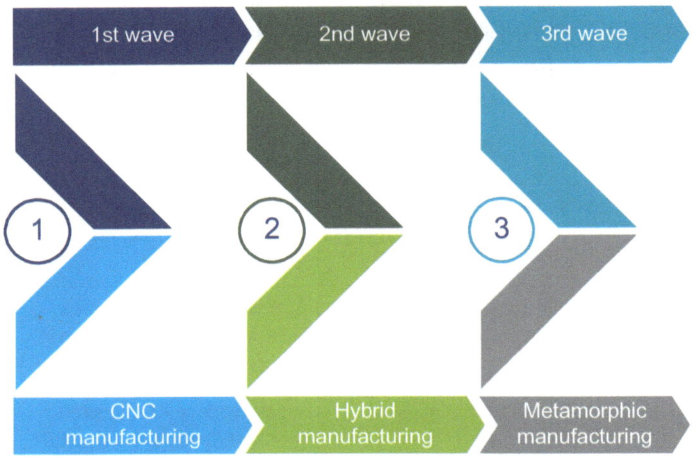

Fig. 1.2: The three digital transformations and the main manufacturing technologies associated.

A key pillar of sustainable manufacturing is the transition from linear to circular production models, emphasizing reuse, remanufacturing, recycling, and responsible sourcing. Products are designed for durability and disassembly, enabling material reintegration and waste reduction. Smart tools support this by tracking energy, resource use, and emissions, aligning operations with circular economy goals.

These efforts are grounded in systems thinking, which evaluates environmental, social, and economic impacts across the product life cycle from raw material extraction to end-of-life [52–54]. Life Cycle Assessment (LCA) identifies environmental hotspots, guiding sustainable design and decision-making. This holistic approach fosters cross-disciplinary collaboration and aligns manufacturing strategies with global sustainability targets, including the UN SDGs.

1.2.1 The First Wave: Computer Control and Design (Approx. 1950s–1990s)

The first wave of digital manufacturing, aligned with the Third Industrial Revolution, marked the transition from manual to automated production through numerical control (NC) and CNC systems (Parsons, 1952 [55–57]). Initially developed post-WWII for aerospace applications, NC machines used punched tape to automate tooling. Key innovations, such as the APT language developed at MIT, laid the groundwork for CNC technologies (Parsons & MIT [58–60]). The advent of transistors and microprocessors enabled direct digital control, significantly enhancing precision and flexibility in subtractive manufacturing (SM).

Parallel advancements in CAD and computer-aided manufacturing (CAM) revolutionized workflows by linking design with machine toolpath generation. Early systems like Sketchpad and DAC-1 evolved from 2D drafting into advanced 3D modeling platforms [59, 61]. The integration of CAD and CAM improved production efficiency, while the introduction of parametric modeling and data standards (e.g., IGES, STEP) enabled interoperability and cross-industry collaboration [61–63]. A representative CNC machine architecture is illustrated in Fig. 1.3.

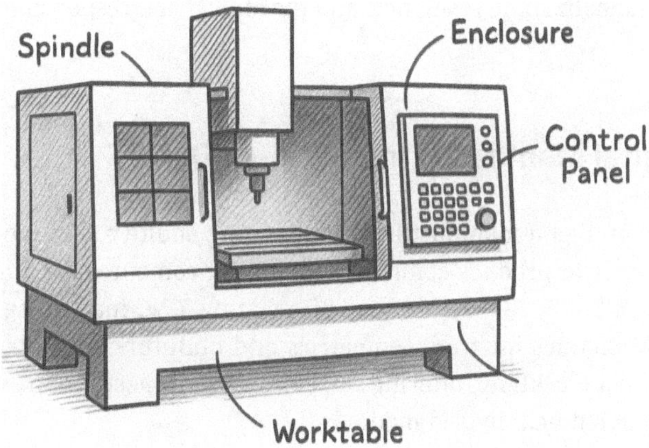

Fig. 1.3: Schematic of a typical CNC machine with the relevant parts.

This wave advanced manufacturing precision, complexity, and productivity, forming the digital backbone for future innovations. However, its subtractive nature led to high material waste and limited control over final properties, which depended on stock materials and post-processing. Nonetheless, it established the essential digital foundation that enabled subsequent manufacturing paradigms.

1.2.2 Second Wave: Additive and Advanced Automation (~1980s–to Date)

The second wave of digital manufacturing, defined by Industry 4.0, marked a shift from basic automation to intelligent, interconnected systems. CPS enabled real-time coordination between computational models and physical operations, enhancing adaptability and precision [11, 64–66]. Key enablers included AM, advanced robotics, and AI, which drove the transition from mass production to mass customization [26, 36, 39, 67].

The AM route evolved from prototyping to producing lightweight, complex components, while robotics advanced toward collaborative autonomy using vision systems and machine learning. The integration of IoT, big data, and cloud computing enabled predictive maintenance, optimization, and agile decision-making [68, 69]. Digital twins further supported real-time simulation and control [70–72], and cloud-connected AM-robotic platforms facilitated decentralized, scalable production [73, 74].

Despite its promise, Industry 4.0 faced material and post-processing challenges in AM [10, 23, 26, 29–31, 38, 39, 75–77], along with implementation barriers across sectors [30, 31]. Still, it fundamentally transformed manufacturing by embedding intelligence, data-centricity, and human-machine collaboration, establishing a foundation for Industry 5.0 and the focus on personalization, resilience, and the ethical and responsible application of AI.

1.2.3 Third Wave of Digital Manufacturing (2016–to Date)

MM represents the third wave of digital manufacturing, integrating additive, subtractive, and transformative processes to produce complex, customized components with enhanced material efficiency [1, 11–13, 78]. By unifying 3D printing, CNC machining, and forming technologies, MM enables intricate geometries and multifunctional designs within a single, streamlined workflow, offering improvements in performance, cost, and sustainability over conventional hybrid methods.

At its core, MM combines digital and physical systems through AI, machine learning, real-time analytics, and robotics to enable predictive maintenance, adaptive control, and decentralized, on-demand production [79–82]. This results in agile, resilient manufacturing environments that respond dynamically to shifting demands.

Emerging capabilities in multimaterial 3D printing, embedded sensors, and gradient materials further expand the potential of MM, fostering sustainable production and innovation in aerospace, healthcare, automotive, and consumer goods. With its ability to deliver mass customization and reduce environmental impact, MM is positioned to transform global manufacturing through intelligent, resource-efficient production.

1.2.4 Comparative Assessment of the Different Waves

Digital manufacturing has evolved through three major waves, as outlined in Tab. 1.2. It has evolved from CNC/CAD/CAM in Wave 1 to AM in Wave 2 and to emerging MM in Wave 3. Each phase introduced new core processes and unique value propositions. Control capabilities have advanced from geometric shaping to microstructural and property-level precision, while material utilization improved toward near-net shaping in Wave 3 [83]. Input materials have evolved from solid blocks to powders and solid preforms, with flexibility evolving from toolpath variation to tailored material properties. These advancements are enabled by digital controls, AI, robotics, and integrated computational materials engineering (ICME), which is underpinned by the rise of intelligent and adaptive manufacturing systems.

Tab. 1.2: Comparative analysis of digital manufacturing waves.

Feature	Wave 1 (CNC/CAD/CAM)	Wave 2 (additive manufacturing)	Wave 3 (metamorphic manufacturing)
Primary process	Subtractive (material removal)	Additive (layer-by-layer application)	Deformative/shaping (incremental thermo-mechanical working)
Material state control	Primarily geometry	Primarily geometry (properties often set by material/process)	Geometry, microstructure, and local properties (integrated control)
Material utilization	Often low (significant waste as chips)	Variable (can be high, but support/powder waste occurs)	Potentially very high (near-net shaping)
Input material	Solid block, casting, and forging	Powder, wire, liquid resin, and filament	Billet, preform, sheet, and plate (solid material)
Property enhancement	Typically, via offline post-processing (e.g., heat treat)	Often via post-processing (e.g., hip and heat treat)	Integrated into primary shaping process
Flexibility type	Geometric complexity and toolpath variation	Geometric complexity and material deposition	Geometric shaping, local property tailoring, and process path variation
Key enablers	Digital control, machine tools, and CAD/CAM	Lasers, extruders, and digital slicing	Robotics, advanced sensors, AI/control, and thermo-mechanics

1.3 Hybrid Manufacturing: Rudimentary Concepts and Advances

HM has emerged as a strategic integration of AM and SM. This is to address demands for efficiency, complexity, and customization [2, 19–21, 84–86]. AM offers geometric freedom through layer-by-layer deposition [26, 36, 38, 39, 67, 87–89], whereas SM provides precision and surface quality via material removal [55–57, 90–92]. HM leverages the strengths of each manufacturing approach often within a single platform, enabling real-time transitions between deposition and machining, thus overcoming the limitations of each individual process [17, 93, 94].

True HM emphasizes synergy rather than sequential operations, enabling intermediate machining during builds and improved part quality. The approach mitigates AM's issues with surface finish and dimensional accuracy, as well as SM's constraints in geometry and material waste. The definitions of HM vary, ranging from tightly integrated single-machine systems to broader multiplatform or energy-assisted workflows [15, 39, 95, 96]. The field distinguishes hybrid processes (e.g., AM + SM) from subhybrid processes (techniques within a single category) and further classifies systems as "assisted" or "mixed," reflecting HM's growing scope and complexity [97, 98].

1.3.1 Key Terminology, Technologies, and Processes of HM, HAM, and HASM

The terminology in HM reflects its rapid evolution and varying degrees of process integration. Configurations range from separate setups to fully integrated systems with simultaneous additive and subtractive interactions (Fig. 1.4), directly influencing complexity and performance [16, 17, 94, 99]. In the absence of universal definitions, understanding the integration depth is essential for evaluating system capabilities.

Generally, HM is defined as the combination of two or more established manufacturing processes within a unified system to achieve synergistic outcomes [17, 94, 97, 99–101]. Thus, hybrid additive manufacturing (HAM) integrates AM with other processes or energy sources to enhance quality and efficiency [97, 100–103]. HM systems incorporate various AM technologies, each with distinct strengths as shown in Fig. 1.5 [16]. Directed energy deposition (DED) is the most prevalent in HM and uses lasers, electron beams, or electric arcs to melt and deposit wire or powder onto substrates, enabling repair, feature addition, and near-net-shape fabrication [16, 17, 83, 94, 97, 99, 100]. Laser metal deposition (LMD), a laser-based DED variant, is successful in precision cladding and for functionally graded materials development when integrated with multiaxis CNC systems. Similarly, the wire arc additive manufacturing (WAAM) employs arc welding methods (e.g., metal inert gas, tungsten inert gas, and plasma arc welding) to deposit metal from wire feedstock in an incremental fashion [104–109]. It offers high deposition rates and low material costs, making it suitable for large components, although it typically requires post-processing due to lower resolution [104–108]. Pulsed arc plasma WAAM improves thermal control, enhancing surface finish and reducing distortion [110–112]. The powder bed fusion (PBF) techniques (selective laser melting and electron beam melting) combine high-resolution AM with in-process milling to achieve tight tolerances and complex internal features, albeit with lower build speeds and volumes. One example is the Matsuura LUMEX system. Generally, a comparative decision matrix summarizing the key AM technologies used in HM is given in Tab. 1.3.

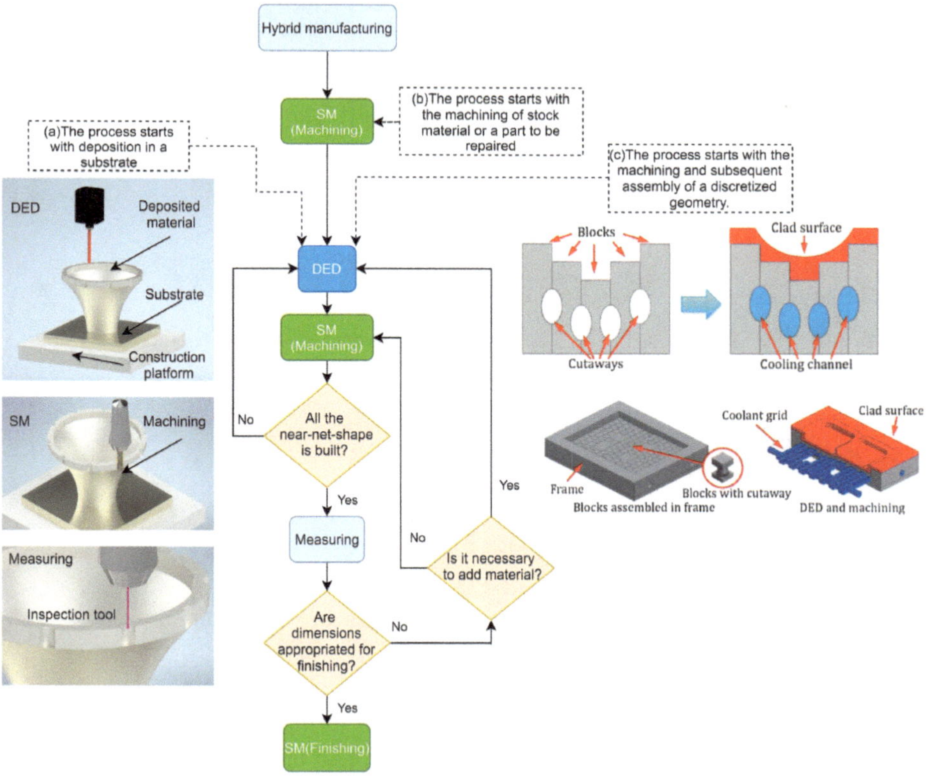

Fig. 1.4: Typical hybrid manufacturing flow diagram (with kind permission of Springer Nature [17] and Elsevier [20]).

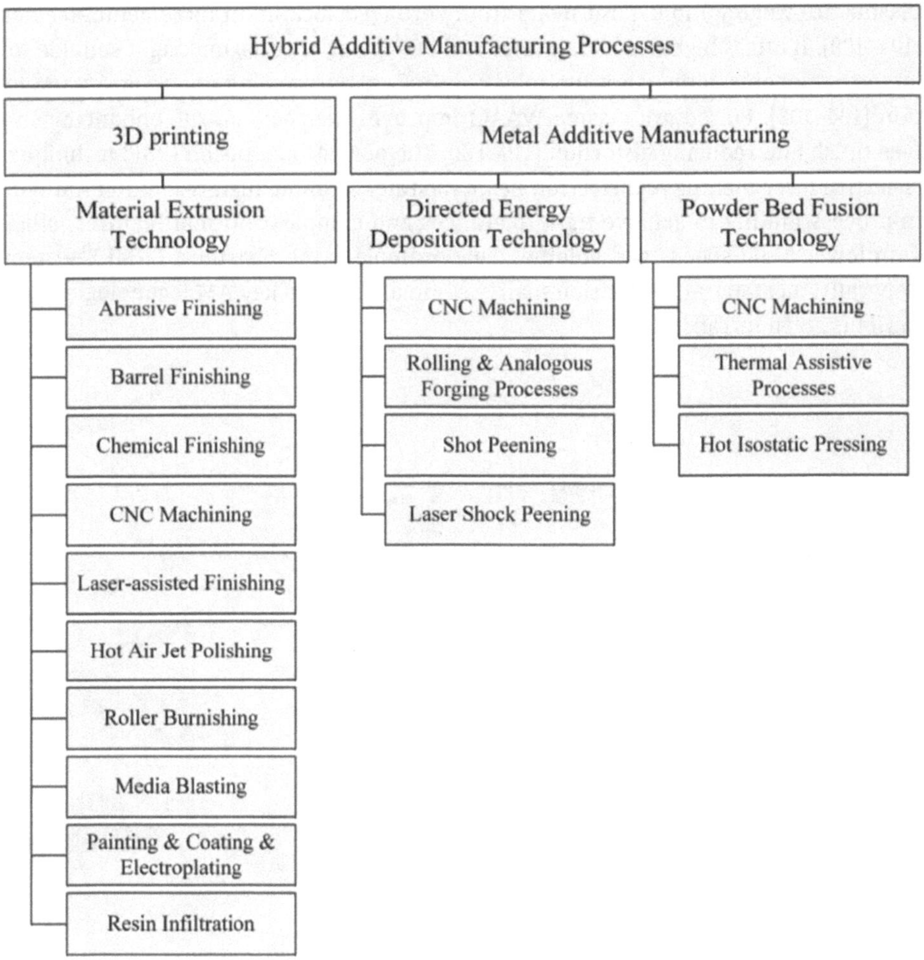

Fig. 1.5: Hybrid process in 3D printing (with kind permission of Springer Nature [16]).

A key subset of HM is the hybrid additive, subtractive, and transformative manufacturing (HASM). This merges AM with machining and represents the most widely adopted form, offering both design freedom and high-precision finishing [16]. As shown in Fig. 1.6, the different HM systems compare the processes based on surface quality, processing time, and processed layer depth. There are subtle trade-offs between precision, time efficiency, and surface improvement across various post-processing techniques (Fig. 1.6).

There are additional AM techniques supporting HM workflows. For instance, cold spray bonds metal powders via supersonic impact without melting, making it suitable for temperature-sensitive repairs and coatings [113]. Furthermore, ultrasonic AM joins metal foils using ultrasonic vibrations and pressure, enabling low-temperature bond-

Tab. 1.3: Comparison of AM technologies in HM systems.

AM technology	Process type	Key strengths	Typical applications	Limitations	Integration in HM
Directed energy deposition (DED)	Powder/ wire + energy beam	Flexible material input, near-net shaping, and repair capabilities	Feature addition, part repair, and structural builds	Moderate resolution and surface finish	Common in CNC-integrated hybrid platforms
Laser metal deposition (LMD)	Laser + powder	High precision, complex paths, and graded material deposition	Cladding, tooling, and multimaterial parts	Cost and requires CNC integration	Multiaxis CNC systems
Wire arc additive manufacturing (WAAM)	Wire + arc welding	High deposition rate and low material cost	Large-scale metal components	Poor resolution and requires significant finishing	Suitable for heavy industry and large parts
Pulsed arc plasma WAAM (PAP-WAAM)	Wire + pulsed arc	Improved thermal control and reduced distortion	Enhanced WAAM applications	Higher equipment complexity	Experimental, for improved quality WAAM
Powder bed fusion (PBF) (e.g., SLM and EBM)	Powder + laser/ beam	High accuracy, fine resolution, and complex internal features	Precision parts and internal channels	Small build volume and slower speed	Integrated with in-process milling (e.g., LUMEX)

ing, dissimilar material integration, and embedded sensors [114, 115]. Material extrusion, typical of FDM processes, is occasionally used in HM for producing jigs, fixtures, or multimaterial parts, though it is less suited for structural metals.

CNC machining forms the core subtractive process in HM. This encompasses milling, turning, drilling, and grinding, which are digitally controlled for precision [16, 17, 94]. Milling shapes complex 3D geometries, turning produces cylindrical parts, drilling creates holes, and grinding ensures fine surface finishes [16]. The CNC systems range from 3-axis to advanced 5-axis configurations, refining AM-produced near-net shapes to meet dimensional and functional specifications [55, 56, 116].

Electrical discharge machining (EDM) is another subtractive technique. It uses controlled sparks to remove material, ideal for hard or delicate components [16, 117]. Dry EDM milling uses gas instead of liquid dielectric to enhance compatibility with WAAM systems through pulsed arc erosion [16, 83, 110, 112]. Wire-EDM is also employed in HM for high-precision cutting, feature generation, and part separation in tight-tolerance applications.

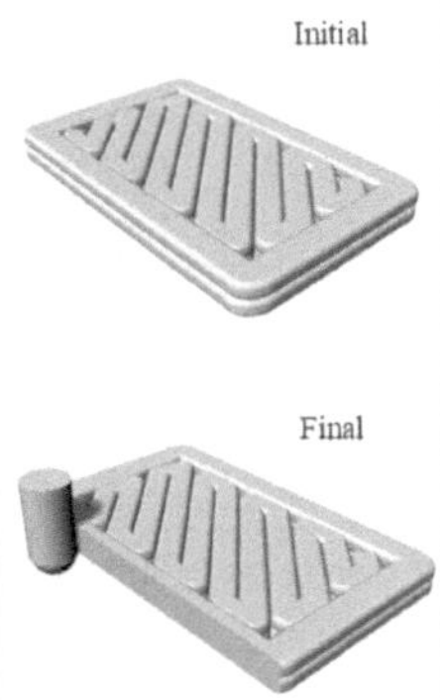

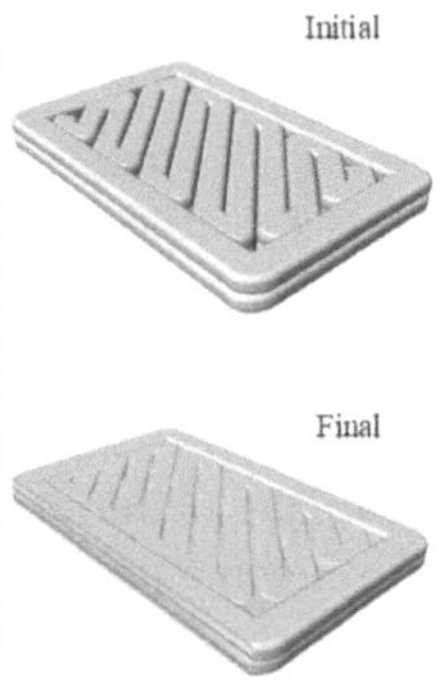

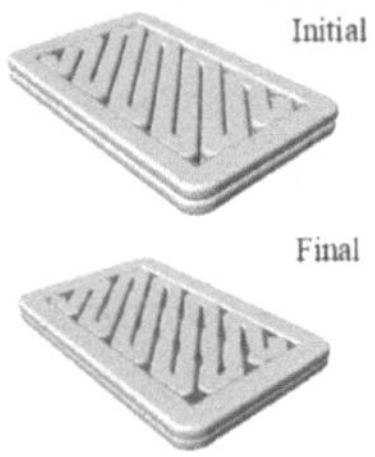

	Secondary Process	Surface Quality Ra [µm]	Processing Time [min]	Processed Layer [µm]
Subtractive Combinations	CNC Machining	2-4	10-42	200-690
	Chemical Finishing	0.1-1	3-40	40-800
	Barrel Finishing	3-5	100-1000	50-400
	Abrasive Finishing	1-6	20-60	50-350
Transformative Combinations	Laser-assisted Finishing	0.3-10	4-10	-
	Hot Air-jet Polishing	0.2-4	3-5	-
	Roller Burnishing	0.11-0.15	2-5	300-350
	Media Blasting	4-12	1-5	-
Additive Combinations	Painting/ Coating	0.7-4	60-3000	60-180
	Resin Infiltration	-	3-240	40-800

Fig. 1.6: Different process combinations for hybrid manufacturing with technical details (with kind permission of Springer Nature [16]).

1.3.2 Integration Platforms and System Architectures in HM

HM systems differ in how additive (AM) and subtractive (SM) processes are integrated, affecting performance, cost, and scalability (Fig. 1.7). Fully integrated machines, such as DMG MORI's LASERTEC and Matsuura's LUMEX, house AM (e.g., DED) and SM (e.g., milling) tools in a single CNC platform, enabling accurate, efficient production without refixturing [16, 17, 94, 118]. Retrofitted systems, like those using Hybrid Manufacturing Technologies' AMBIT™ heads, offer a cost-effective alternative by upgrading existing CNC machines, though with less seamless integration [17, 94, 97, 99, 100, 115].

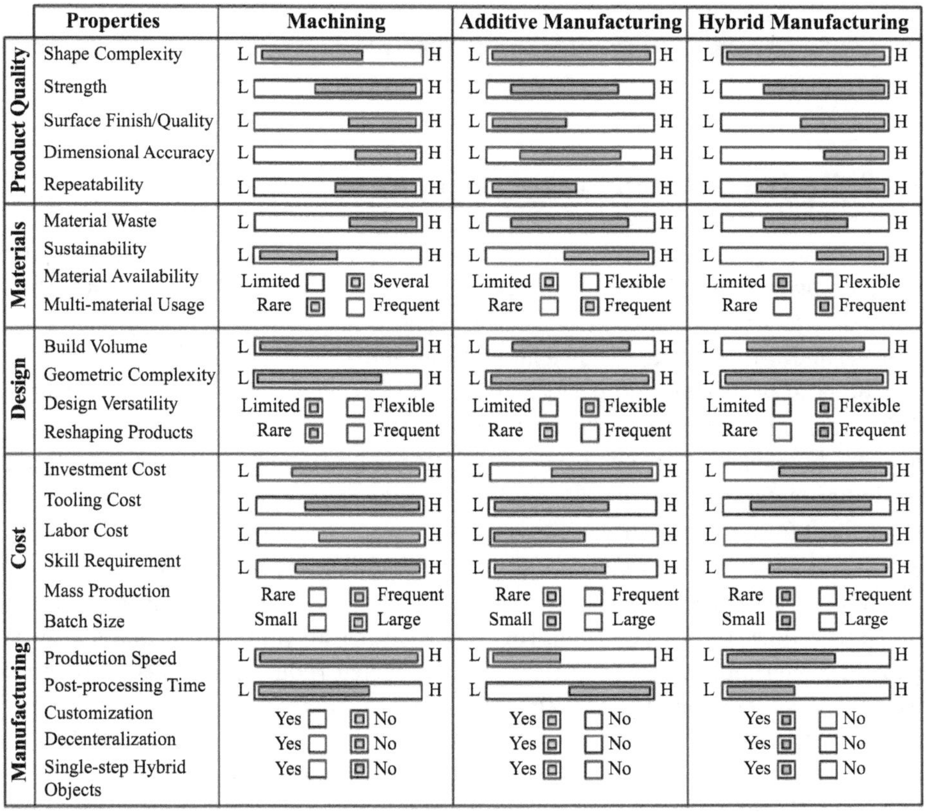

Fig. 1.7: Comparison of machining, additive, and hybrid manufacturing based on critical parameters (with kind permission of Springer Nature [16]).

Modular robotic systems use AM-enabled robotic arms alongside separate or integrated machining stations, providing flexibility for large or complex parts, though they increase coordination complexity and alignment risk [34, 119, 120]. Across all architectures, advanced process planning software is essential for toolpath optimization, thermal simulation, and collision avoidance [16, 49, 72]. However, a lack of mature, user-friendly HM software remains a key challenge [16].

Each architecture involves trade-offs: integrated machines offer precision but require high investment; retrofits are cost-effective but less optimized; robotic systems offer flexibility at the cost of complexity [16]. Selection depends on factors such as part geometry, materials, production scale, and infrastructure.

HM combines the strengths of machining and AM, offering a balanced, high-performance solution. A comparative analysis across product quality, materials, design, cost, and performance (Fig. 1.7) shows that machining produces the best surface finish and accuracy but lacks flexibility and sustainability [16, 121]. AM enables design freedom and material efficiency but often requires post-processing and may compromise mechanical strength. HM addresses these gaps, achieving high complexity, strength, and precision while enhancing flexibility and reducing waste [122, 123], making it a compelling strategy for advanced, efficient production.

1.3.3 Strategic Advantages of HM Processes

Design freedom and complexity: The process combines AM and SM processes to enable complex geometries, internal features, and lightweight structures that are difficult or impossible to achieve with SM alone [16]. It supports part consolidation and allows precise machining during intermediate build stages, enhancing functional integration [16].

Material properties and performance: HM improves surface finish and reduces defects typical in AM, enhancing fatigue life, wear resistance, and dimensional stability [122]. It enables tailored material properties via functionally graded materials and dissimilar metal cladding [16, 124], with some systems incorporating in-process grain refinement for improved strength.

Lead time and cycle reduction: By integrating multiple steps of AM, SM, and inspections into a single platform, HM reduces setup times and part transfers, streamlining workflows [16, 17, 23, 24, 94]. This accelerates prototyping and maintenance, repair, and overhaul (MRO) operations, with reported reductions in lead time from weeks to days across industries such as aerospace and defense.

Cost efficiency and material utilization: HM minimizes material waste through near-net-shape builds, significantly reducing costs for high-value materials such as titanium [125]. Case studies report material loss reductions of up to 97% (e.g., WAAM vs. milling), and cost savings up to 67% for specific parts and tooling. Retrofitting offers a low-cost HM entry point.

Accuracy and surface quality: By integrating precision machining, HM ensures tight tolerances and high-quality finishes, critical for fluid flow, sealing, and enhanced mechanical properties (wear resistance and fatigue strength) [16]. Internal features can be machined midbuild, achieving levels of accuracy and surface quality not possible with AM alone.

1.3.4 Comparative Analysis: Hybrid Manufacturing with Standalone Processes

HM offers a balanced solution that overcomes key limitations of AM and SM. Each approach possesses distinct strengths and weaknesses, and HM aims to find a synergistic balance. The comparative analyses are summarized based on specific features in Tab. 1.4.

HM versus AM: HM outperforms standalone AM in part quality, precision, and surface finish. By integrating machining, HM achieves tighter tolerances, improved aesthetics, and reduced residual stresses particularly for internal features inaccessible post-build. However, HM incurs higher equipment costs and operational complexity, requiring advanced planning and skilled operation. AM is excellent for geometric freedom and low initial costs, making it ideal for complex shapes without toolpath constraints. Yet, it typically has a drawback of surface roughness, dimensional inaccuracy, and internal defects, often requiring extensive post-processing, which HM can streamline.

HM versus SM: Compared to SM, HM enables intricate geometries, internal features, and near-net-shape builds with less material waste. It supports repair, part consolidation, and multimaterial fabrication, offering advantages in prototyping and customized production. SM remains superior for simple geometries, high throughput, and consistent material properties, benefiting from process maturity and standardization. However, it is limited in handling complex designs and is less material efficient. HM bridges these gaps, though it is better suited for low- to medium-volume, high-value applications due to its higher capital and process requirements.

Tab. 1.4: Comparative matrix of subtractive and additive manufacturing technologies with hybrid models focusing on unique features.

Feature	Hybrid manufacturing (HM)	Additive manufacturing (AM)	Subtractive manufacturing (SM)
Geometric complexity	Very high (combines AM freedom with SM refinement)	Highest (least constrained initially)	Medium to high (limited by tool access, especially internal)
Internal feature capability	High (AM builds, SM can finish during build)	High (built-in, but often poor internal finish/accuracy)	Very low (limited by line-of-sight tool access)
Achievable accuracy/ tolerance	High (limited by SM capability)	Low to medium (process dependent and often requires post-processing)	Very high (process dependent, capable of tightest tolerances)
Surface finish quality	High (finished by SM)	Low to medium (inherent layering and requires post-processing)	Very high (smooth finishes achievable)

Tab. 1.4 (continued)

Feature	Hybrid manufacturing (HM)	Additive manufacturing (AM)	Subtractive manufacturing (SM)
Material waste (buy-to-fly)	Low (AM builds near-net shape)	Lowest (material added only where needed)	High (significant material removal from stock)
Material variety/ availability	Medium to high (growing, depends on specific AM process)	Medium (more limited than SM and process specific)	Very high (widest range of metals, plastics, composites, etc.)
Speed (prototyping)	Very fast (for complex and high-fidelity prototypes)	Fast (for initial shapes, less so for finished quality)	Medium to slow (requires setup, programming, and machining time)
Speed (production volume)	Low to medium (often limited by AM deposition rate)	Low (generally slowest for volume production)	High (can be very fast for optimized processes)
Cost (low volume/ custom)	Medium to high (high equipment cost, but avoids tooling)	Low to medium (lower entry cost and avoids tooling)	High (tooling/setup costs dominate low volumes)
Cost (high volume)	High (process time or operational costs can be high)	Very high (slow speed and material costs)	Low (tooling amortized and fast cycle times)
Repair or modification capability	Very high (ideal for adding material precisely)	Medium (can add material, but integration/finish issues)	Low (difficult to add material back)
Required skill level	High (requires combined AM, SM, materials, and software skills)	Medium (requires AM process knowledge and software skills)	Medium to high (requires machining and programming skills)

1.3.5 Applications and Case Studies Across Industries

HM is gaining traction in sectors requiring precision, complexity, customization, and repair [16]. Early adoption focuses on high-value use cases where ROI justifies investment, paving the way for broader applications [16]. Sector-specific benefits of HM technologies are summarized in Tab. 1.5.

Aerospace and defense: HM is widely used for producing lightweight, high-strength components (e.g., turbine blades and fuel injectors) with internal features and tight tolerances [16]. In MRO, DED or LMD enables efficient repair of costly parts like Inconel blades, reducing downtime and costs [26, 39, 87]. Systems such as those used aboard the USS Bataan support decentralized, on-demand repair, improving fleet readiness. Selected aerospace case studies are presented below.

Tab. 1.5: Sector-specific benefits of HM technologies with the various driving metrics.

	High performance part and accessories	Customization and repairs	Reduced lead time and overall cost
Aerospace and defense	√	√	√
Automotive	√	√	√
Medical and biomedical devices	√	√	√
Tooling, molds, and dies		√	√
Maintenance, repairs, and overhaul			√
Emerging applications	√	√	√

GE turbine blade repair: Using Hybrid Manufacturing Technologies' AMBIT™ system, GE applied DED and machining to restore worn turbine blades. The process enabled precise material addition and remachining to original specifications, eliminating the need for full replacement, reducing repair steps, and enabling performance upgrades yielding multimillion-dollar annual savings.

Boeing winglet production: Boeing reduced the cost of 737 winglet components from $60,000 to $20,000 and lead times from 6 weeks to 4 days by implementing HAM solutions. This approach enhanced production autonomy and supply chain resilience with ~65% cost and ~80% time savings.

TiAl Aero-engine component repair: Hybrid DED-milling techniques were used to repair titanium aluminide turbine blades, offering a viable alternative to costly replacements [126, 127]. Although specific figures were not disclosed, similar MRO implementations (e.g., M1 Abrams turbine parts) have demonstrated savings of ~US $5 million annually [128].

Airbus A350 XWB spare parts: Airbus adopted HAM with subtractive finishing to produce A350 XWB spare parts on demand. This strategy reduced inventory requirements, minimized aircraft-on-ground risk, and improved fleet readiness while lowering operational costs.

USS Bataan – onboard HM: The USS Bataan became the first U.S. Navy ship to deploy a Phillips Additive Hybrid system, integrating metal 3D printing with CNC machining. The system enabled fabrication of a critical sprayer plate in 5 days, dramatically faster than traditional supply methods, enhancing operational readiness and onboard self-sufficiency.

Babcock and AMFG – digital defense supply chain: Babcock partnered with AMFG to develop a hybrid-enabled digital aftermarket supply chain for military logistics. The platform integrates AM with supply chain software to streamline part identification, production, and delivery, reducing reliance on conventional inventory and improving fleet readiness through distributed manufacturing.

Automotive: In automotive, HM accelerates prototyping and produces lightweight, high-performance parts. It also supports rapid tooling, enabling faster production cycles and greater flexibility for jigs, fixtures, and custom components [16, 17, 94]. Some selected case studies are presented below.

Ford GT prototyping: During development of the Ford GT, HM enabled rapid prototyping with integrated additive and subtractive processes. This approach accelerated design iteration, reduced lead times, and enhanced collaboration through quicker production of functional prototypes.

Volkswagen Tiguan components: Volkswagen utilized AM with hybrid finishing to produce select parts for the Tiguan model, achieving faster turnaround and lower production costs without compromising quality, thereby enhancing order fulfillment efficiency.

Accelerated prototyping (general): Across industries, HM streamlined prototyping by unifying AM and SM into a single workflow. This integration enabled rapid part creation and precision finishing, reducing design cycles from over 6 months to under 2 weeks, and supporting faster iteration and product validation.

Medical and biomedical devices: HM enables the production of patient-specific implants with porous geometries for osseointegration and precise finishing for biocompatibility [129–131]. It also supports efficient dental workflows and surgical tool manufacturing, with demonstrated reductions in cost and lead time [132–136]. As HM technology continues to develop with collaboration between industries and academia, various strategic outputs can be realized. For instance, Datron, in collaboration with Concept Laser and Follow Me Technology Group, integrated additive and CNC milling processes to execute hybrid workflows through smart networking [122]. Key innovations include transferring the zero point via infrared touch probes and measuring pins built during sintering, ensuring high-precision post-processing. This seamless integration eliminates the need for recalculating CAM data and maintains object stability, particularly for implant superstructures. Their approach reduces manufacturing costs by 30–50%, demonstrating significant impact on efficiency and precision in medical device production. Some case studies are presented.

Patient-specific implants: HM enables the production of customized orthopedic and dental implants by combining AM (e.g., PBF and DED) for anatomical and porous geometries with SM for high-precision surface finishing. This approach improves fit, function, and cost-efficiency over conventional methods.

High-precision medical component fabrication: A complex metal device was manufactured using metal AM, followed by wire-EDM guided by built-in Datum Reference Features for accurate alignment. The hybrid process achieved <1% dimensional deviation and a 6:1 buy-to-fly ratio, demonstrating precision and material efficiency using separate AM and SM platforms.

Tooling, molds, and dies: HM enhances tooling with features like conformal cooling channels, improving cycle times and part quality. Platforms like Matsuura LUMEX show cost reductions of up to 67%. HM also enables selective repair of worn tools, extending service life and minimizing waste [99]. Some case studies are presented.

Matsuura LUMEX molds: The LUMEX system integrates selective laser sintering with high-speed milling to produce injection molds featuring complex conformal cooling channels. This hybrid process reduces lead times and manufacturing costs by 50–67% compared to conventional methods [99], while enhancing part quality and molding efficiency.

ArcHLM molds: Arc hybrid-layered manufacturing combines WAAM with CNC milling after each layer or set of layers. This approach delivers cost-effective, rapid production of prototype molds and tools with improved surface finish and interlayer bonding, all on a single platform.

MRO: HM is critical in MRO across aerospace, defense, and energy. It restores mission-critical components with precision, reducing reliance on spare inventories and lowering life cycle costs [17, 98, 99]. Notably, the U.S. Army reported saving $5 million annually by using HM to refurbish turbine parts [137].

Emerging applications: HM is expanding into energy (turbine repair), construction (custom nodes), mining (wear-resistant tools), rail (precision parts), consumer goods (customized sports gear), and electronics (embedded cooling or conductive features). Its adaptability and material efficiency support innovation across diverse sectors.

1.3.6 Challenges and Limitations of Hybrid Manufacturing

Economic barriers: High capital costs are required for integrated systems combining AM and CNC. Furthermore, materials are expensive, and energy consumption and maintenance are high, hence a major adoption hurdle. While HM is cost-effective for low-volume, complex, or high-value parts, HM struggles to compete with mass production methods for simple and high-throughput applications.

Technical complexity: Merging AM and SM introduces integration challenges requiring precise alignment, advanced control systems, and intelligent software. Maintaining coordinate accuracy and generating adaptive toolpaths remain significant issues. The lack of standardized, user-friendly process planning tools continues to impede scalability and ease of adoption.

Materials limitations: Material compatibility, residual stress, and thermal distortion are critical concerns. Multimaterial builds demand careful control to avoid defects and ensure reliability. The limited availability of qualified HM materials and challenges in powder handling, safety, and recycling add to operational complexity.

Workforce skill gaps: HM demands cross-disciplinary expertise in CNC, AM, materials science, robotics, and digital manufacturing. The current talent shortage and need for specialized training programs hinder widespread implementation, requiring investment in workforce development and organizational adaptation.

Scalability constraints: HM excels in customized, high-value applications but is limited in mass production due to slower build rates, higher per-part costs, and size constraints. For broader scalability, HM must offer value that outweighs these drawbacks, typically in sectors like aerospace MRO, medical devices, and tooling, where complexity and responsiveness are critical.

1.3.7 Strategic Guidelines for HM Adoption

The adoption of HM requires a strategic and phased approach that commences with application-centric evaluation. This is to ensure that opportunities are assessed on both the part cost and the total value factors, such as complexity, lead time, material savings, inventory reduction, and design freedom, to target applications where HM offers clear advantages. To align with current technological trends, selecting appropriate HM architectures, whether integrated machines, retrofits, or robotic cells with other AM/SM technologies such as DED, PBF, WAAM, CNC, or EDM, should match specific requirements, budgets, and infrastructure, with prioritized seamless integration. Moreover, robust process planning must be prioritized through investments in hybrid-capable CAD/CAM/CNC software, effective coordinate management, advanced thermal control, and hybrid-specific quality assurance to guarantee accuracy and efficiency. In parallel, workforce development should be a key strategic pathway that should be explored. This plays a vital role in ensuring cross-disciplinary training across AM, SM, materials, software, and robotics, and the fostering of collaboration between design, manufacturing, and quality control teams. Thus, phased implementation is advisable, starting with pilot projects in MRO, tooling, or complex prototyping, with retrofitting existing equipment serving as a cost-effective entry point to mitigate risks. Finally, ecosystem engagement ensures long-term success by fostering collaborations with vendors, researchers, and industry groups to access innovations, best practices, and the wider support networks essential for scaling HM adoption.

1.4 Metamorphic Manufacturing (MM): New Frontier and Its Advances

MM is an emerging digital paradigm that combines incremental, thermo-mechanical deformation with real-time control to shape materials and engineer their internal properties. Based on a 2019 study commissioned by TMS (The Mineral, Metals and Materials Society), a definition was coined [18]. The approach is often referred to as "robotic blacksmithing" [11, 18]. Unlike prior digital manufacturing waves focused solely on geometry, MM prioritizes property control through the integration of materials science, sensing, and closed-loop feedback [11, 18]. It operates by applying localized actions such as bending, squeezing, or hammering, coupled with targeted heating, to control grain structure, phase changes, and mechanical performance. This thermomechanical synergy aligns with ICME principles, enabling simultaneous shaping and material enhancement. Key to MM is its die-less, sensor-driven robotic system, which uses real-time data and AI to adaptively adjust tool paths and process parameters. This flexibility supports efficient, low-volume production and customized parts. In addition to its standalone benefits such as high material efficiency and near-net shaping, MM can enhance hybrid workflows by refining AM preforms in both geometry and microstructure. Merging the strength of forging with digital precision, MM offers a transformative path toward intelligent, adaptive manufacturing.

1.4.1 Metamorphic Manufacturing: Key Concept and Enabling Technologies

MM introduces a third paradigm in digital manufacturing, which is thermomechanical deformation that simultaneously shapes materials and tailors their internal properties. Unlike subtractive or additive methods, MM morphs material form and function in tandem using AI, sensors, and robotics to control localized, incremental deformation and thermal inputs. It enables site-specific property control, supporting functionally graded materials without fixed dies [1, 11–13, 78]. It combines the mechanical performance of forging with the flexibility of digital design. Its near-net shaping minimizes waste and energy use, making it a sustainable alternative [11, 18]. Rather than replacing AM or SM, MM complements them within emerging hybrid platforms like smart manufacturing multiplex systems. Table 1.6 discusses enabling technologies for MM. Robotics and advanced actuators provide precise deformation, whereas sensing systems provide real-time feedback. Thermal management, adaptive control, and AI close the loop for process optimization [1, 11–13, 78]. The merits and challenges are summarized in Tab. 1.7 with emphasis on sensor integration, real-time analytics, and computational modeling costs, requiring robust solutions. This reflects that the technology remains work in progress with the need for further research.

There are key elements and enabling technologies driving MM mechanisms. These, summarized in Fig. 1.8, include precision robotics, adaptive actuators (e.g., presses and rollers), and reconfigurable tooling for die-less forming. An integrated sensor suite comprising optical, thermal, and force sensors provides real-time feedback for closed-loop control [1, 11–13, 78]. Localized thermal management (via induction, laser, or plasma) synchronizes with deformation to guide phase transformations and microstructural

Tab. 1.6: Key enabling technologies for MM.

Category	**Specific examples**	**Function in MM**	**Current maturity/ challenges**
Robotics	High-precision, multiaxis heavy-duty robotic systems	Tool/workpiece with controlled motion and force application	Mature technology, but demands high stiffness and precision; real-time adaptive control remains complex.
Actuators and forming tools	Servo-driven presses, hammers, rollers, and adaptable tooling systems	Apply controlled incremental deformation forces	Mature servo control; high-force, precision actuators available; universal tooling still in development.
Sensing (metrology and thermal)	Structured light, laser scanning, vision systems, IR cameras, and pyrometers	Real-time 3D shape and surface temperature sensing for closed-loop geometric and thermal control	Mature sensors exist, but integration, speed, accuracy, and robustness in harsh environments remain key challenges, especially calibration, emissivity, and resolution.
Sensing (force/ strain/acoustic emission (AE))	Force, vibration, and acoustic sensors (e.g., load cells, accelerometers, and ultrasonics)	Monitor forces, assess material behavior, and detect defects or anomalies	Sensors are available, but real-time interpretation of complex signals (e.g., AE, ultrasonics) demands advanced analytics.
Thermal management	Induction, laser, plasma, or flame heating; gas/ liquid jet cooling systems	Precision thermal control synchronized with deformation	Existing technologies require precise timing and integration; rapid, localized thermal cycling remains challenging.
Adaptive control and AI	AI/ML-driven real-time feedback control loops	Integrate sensor data to dynamically adjust parameters, optimize performance	Control theory is mature, but effective AI/ML use requires extensive data, robust algorithms, and validation for safety-critical applications
Computational modeling (ICME)	FEA-based simulations (deformation, thermal, and microstructure) and data-driven models	Process planning, shape/ property prediction, and control optimization	Significant progress made, but accuracy depends on material data and model validation; high computational cost remains a challenge.

Tab. 1.7: Summary of metamorphic manufacturing advantages and challenges.

Category	Specific point	Supporting evidence/explanation
Advantage	Flexibility and customization	Die-less nature enables diverse geometries, small batches, and rapid changes.
	Efficiency (potential)	Faster time-to-market vs. forging (no tooling delay), potential for faster cycles versus some AM, and reduced setup in integrated systems.
	Enhanced/tailored material properties	Integrated thermo-mechanical control allows microstructure tailoring for superior/localized properties (strength and toughness).
	Waste reduction	Near-net shaping minimizes material scrap compared to subtractive methods.
	Energy efficiency (potential)	Deformation can be less energy-intensive than melting (AM) or extensive machining.
	Reduced tooling costs and lead times	Eliminates/minimizes need for expensive, part-specific dies, crucial for low-volume/prototyping.
Challenge	Process complexity	Controlling coupled deformation, thermal effects, and material response is highly challenging.
	Cost (initial investment and operation)	High cost of advanced robotics, sensors, control systems, software, computation; achieving cost-competitiveness is difficult.
	Standardization	Lack of established standards for processes, data, quality assurance, materials hinders adoption.
	Scalability	Scaling to large parts or high-volume production with technical economic hurdles.
	Process control and quality assurance	Ensuring consistency requires robust real-time monitoring, NDE, adaptive control; managing distortion/residual stress is key.
	ICME integration and accuracy	Reliance on accurate, validated material models and sufficient computation; model development is costly/complex.
	Workforce skills and technological maturity	Requires interdisciplinary expertise in materials, robotics, AI, modeling, etc. Integration of enabling technologies into robust industrial systems is ongoing R&D.

evolution. The intelligence of MM lies in advanced control systems and AI-driven predictive models that interpret sensor data and dynamically adjust processing parameters [11, 18]. The ICME frameworks support this by modeling heat flow, deformation, and material behavior. However, integration of these complex components into a cohesive, scalable system remains a major challenge for industrial deployment.

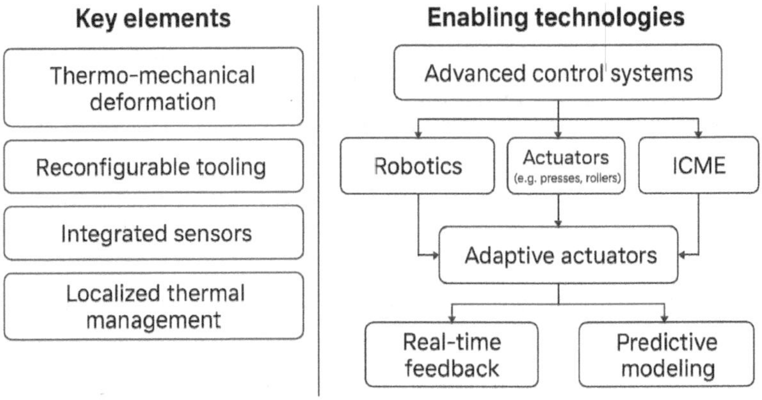

Fig. 1.8: Key elements and enabling technologies driving MM evolution.

The digital reshaping process or robotic blacksmithing relies on six core components. These elements combine traditional blacksmithing techniques with advanced digital technologies to enable precise, adaptive, and intelligent manufacturing, as shown in Fig. 1.9. The sensing components enable real-time monitoring. Advanced sensors track key parameters, such as temperature, shape, and internal structure of the metal, as it is being formed [1, 11–13, 78]. This real-time data provides the necessary input for precise control of the shaping process. Thermal controls are used to adjust material properties from phase transformation perspectives. Localized heating methods like lasers or induction are applied to make the metal more workable by inducing the desired phase change [11, 18]. By targeting specific areas, manufacturers can achieve complex shapes and control material characteristics with high precision. Actuation and forming tools perform the deformation. Robotic systems equipped with interchangeable tools such as hammers and presses incrementally shape the metal. This mirrors traditional blacksmithing but with greater consistency, speed, and accuracy.

Robotic manipulation ensures precise handling. Robotic arms and positioning platforms move and orient the workpiece throughout the process. This ensures each deformation step is carried out with uniformity and spatial precision. Computation guides the forming process. AI and advanced simulations analyze sensor data and predict how the material will respond [138]. These tools allow for real-time adaptation and optimization of the process. Furthermore, the closed-loop feedback control ensures accuracy. By comparing actual results with desired specifications, the system makes dynamic adjustments [1, 11–13]. This continuous feedback loop guarantees that the final product meets exacting standards. These components transform blacksmithing into a digitally driven, high-performance manufacturing approach.

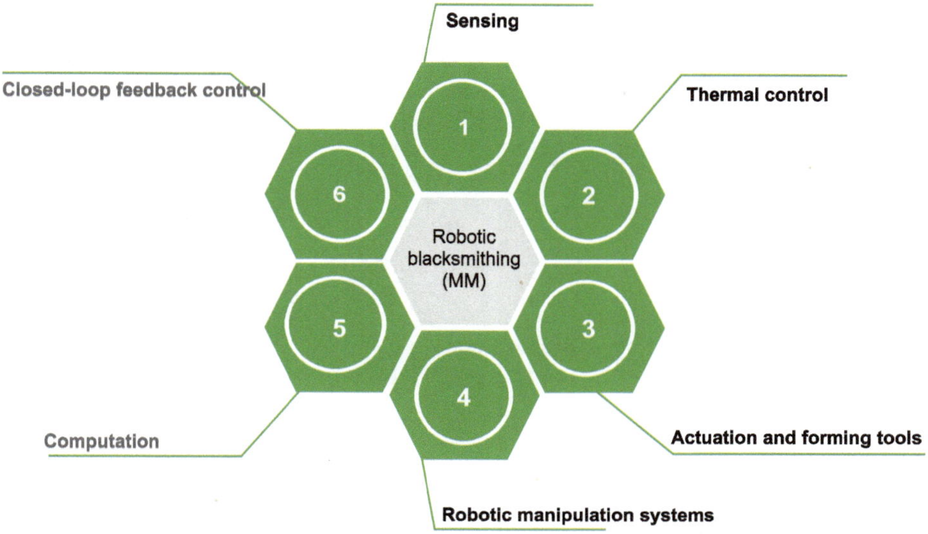

Fig. 1.9: Six critical components of metamorphic manufacturing.

1.4.2 Broader Impact of MM

A schematic showing the strategic broader impact matrix titled of MM is summarized in Fig. 1.10. This is based on the transformative potential of MM across four key dimensions, which are the on-demand production, mass customization, supply chain resilience/localization, and sustainability [11, 18]. On-demand production is positioned as high in both feasibility and impact, emphasizing MM's die-less, digital responsiveness to urgent or low-volume manufacturing needs. Mass customization also scores highly, reflecting the ability of MM to economically vary geometry and material properties without tooling changes, which is ideal for personalized or niche applications [1, 11–13, 78]. Supply chain resilience and localization rank high in strategic value but slightly lower in current maturity, highlighting the potential to decentralize production and mitigate global disruptions. Sustainability is shown as a cross-cutting advantage, offering environmental and cost benefits via reduced waste and energy use, thus aligning with global decarbonization efforts [10, 139, 140]. Thus, Fig. 1.10 positions MM as a key enabler of future-ready, agile, smart, and sustainable manufacturing systems.

Based on the previously discussed merits, MM further distinguishes itself through its on-demand production capabilities. By enabling rapid, die-less fabrication of parts directly from digital models, MM significantly reduces lead times, simplifies logistics, and supports just-in-time manufacturing strategies that minimize inventory costs while improving responsiveness [1, 11–13, 78]. In addition, MM facilitates true mass customization. Its programmable, die-less forming processes make small-batch and individualized production economically viable [1, 11–13, 78], a feature especially valu-

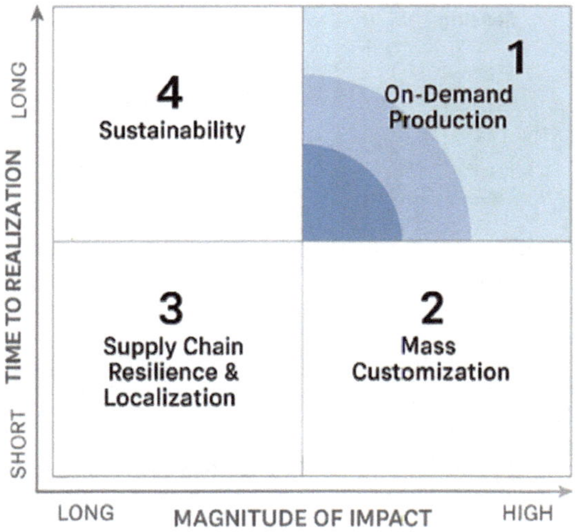

Fig. 1.10: Broad impact of MM based on four critical dimensions that drive the manufacturing sector.

able in sectors such as medical devices and consumer goods, where both geometric flexibility and tailored material properties are critical without sacrificing performance or scalability. Furthermore, from an economic standpoint, MM enhances supply chain resilience and promotes localized or decentralized manufacturing. With minimal infrastructure requirements, it supports decentralized production models that reduce dependency on global networks, a capability essentially valuable in defense, remote operations, or disaster-response contexts. Finally, MM advances sustainability by functioning as a near-net shaping process that conserves both material and energy, offering a greener and more cost-effective alternative to subtractive or melting-based additive methods, fully aligned with circular economy principles.

1.4.3 Potential Application Sectors for Metamorphic Manufacturing

MM, though still emerging, holds substantial promise for industries that demand high-performance, customizable, and agile production solutions [1, 11–13, 78]. By enabling simultaneous material shaping and in situ property control, MM offers capabilities that surpass traditional additive and subtractive methods, making it especially valuable in scenarios where geometric complexity, mechanical performance, and adaptability are essential.

1.4.3.1 Potential Application Sectors

In the aerospace sector, MM is well suited to producing high-strength, lightweight components such as engine mounts, landing gear, and structural airframe parts. It also offers a viable path for replacing legacy components without original tooling. Ongoing research into MM's application with advanced alloys like Al-Li and high-strength aluminum supports industry efforts to reduce weight and accelerate design iteration cycles [1, 11–13, 78].

In the automotive industry, MM addresses challenges related to forming complex shapes from advanced alloys, particularly in prototyping, niche models, and low-volume production [11, 18, 141–144]. Its die-less, programmable approach eliminates the need for expensive stamping tools, offering a cost-effective alternative for manufacturing structurally demanding components with improved material efficiency.

The defense and naval sectors stand to benefit significantly from the on-demand production capabilities of MM [11, 18]. By reducing dependence on global supply chains and enabling localized part manufacturing, especially for mission-critical systems like submarines, armored vehicles, and naval vessels, MM strengthens supply chain resilience. It has garnered attention from institutions such as the Office of Naval Research and defense consortia like LIFT, underscoring its strategic relevance [1, 11–13, 78].

In healthcare and biomedical applications, MM enables the production of patient-specific implants using biocompatible alloys. Its ability to tailor both geometry and mechanical properties, such as stress shielding and fatigue resistance, enhances implant integration and longevity, particularly in orthopedic and dental domains [11, 13, 18].

The energy sector may leverage MM for fabricating components exposed to extreme thermal and mechanical conditions, such as turbine blades or reactor internals. The microstructural control in MM allows for the customization of properties to meet rigorous operational demands, making it valuable in both traditional and next-generation energy systems [1, 11–13].

Finally, MM shows strong potential in tooling applications, particularly when integrated with AM. This hybrid approach can be used to build and then shape or repair high-performance dies and molds, offering improved surface finish and mechanical robustness [18]. Such capabilities are ideal for specialized tooling, where precision, durability, and efficient turnaround are critical.

1.4.3.2 Case Studies and Recent Advances

MM remains in its early stages but is gaining momentum through focused research and institutional support. For instance, LIFT, a U.S. Manufacturing Innovation Institute, has championed MM under its "Novel/Agile Processing" agenda, funding key initiatives such as the Robotic Blacksmithing Challenge to demonstrate feasibility and promote workforce engagement [11, 12, 18]. These efforts underscore MM's potential to unify geometric shaping with in-situ material property control.

Academic institutions are central to MM's development. Ohio State University, led by Professor Glenn Daehn, has pioneered robotic incremental forming through real-time feedback, robotics, and localized thermal control [11, 18]. His leadership in the HAMMER Engineering Research Center aims to scale MM architectures for industrial relevance [1, 11–13, 78]. Similarly, Professor Jian Cao at Northwestern University has advanced hybrid and deformation-based processing, emphasizing integrated modeling and adaptive tooling.

Research is also targeting the digital infrastructure required for MM's industrialization, blending materials science, mechanical engineering, and intelligent automation. Demonstration platforms like the smart manufacturing multiplex, which combines DED, machining [55, 56], and sensing [145–147], illustrate the hybrid convergence central to MM and highlight challenges in system integration.

Defense organizations, particularly the Office of Naval Research, are showing strong interest due to MM's alignment with localized, on-demand manufacturing and its suitability for producing near-net-shape components in constrained environments [18]. Given these factors, MM is likely to see early adoption in defense and aerospace applications, with broader uptake dependent on progress in cost reduction, reliability, and process standardization.

1.5 Conclusion and Future Directions in Digital and Metamorphic Manufacturing

MM is a significant evolution in digital manufacturing, emerging as a distinct third paradigm from subtractive CNC machining and AM. Built on incremental thermomechanical deformation and digitally guided processes, it integrates robotics, AI, sensing, and integrated computational materials engineering to control geometry and material properties instantaneously. The technique enables excellent design flexibility, waste reduction, energy efficiency, and enhanced sustainability, thus, positioning MM as an innovative pathway for advanced production. Furthermore, a defining feature of MM is its die-less capability, which supports on-demand production and mass customization of complex components. However, widespread adoption is still limited by major challenges such as high capital investment, scalability, lack of standards, quality assurance, and the development of advanced control systems. Addressing these barriers will be critical if MM is to achieve industrial-scale integration.

A central research priority is the integration of constituent technologies. Although sensing, robotic manipulation, thermal control, and computational modeling have advanced individually, their operation is in isolation rather than being integrated. Future research progress will depend on establishing standardized, interoperable architectures for seamless communication across sensors, actuators, and control software for adaptive and closed-loop manufacturing platforms.

Standardization and certification present another challenge. In contrast to CNC and AM, which have evolved with established qualification protocols, MM is still at its infancy without consistent taxonomies, standards, and certification frameworks. This will be challenging for MM to be adopted in highly regulated sectors such as aerospace and defense. Hence, the need to develop robust qualification pathways will be critical for industry trust, safety, compliance, and reliability. Furthermore, the advancement of modeling and simulation is a gap that require in depth and further research engagement in ways that is rational. For instance, effective MM requires lightweight, high-fidelity predictive models and digital twins capable of capturing complex behaviors such as thermal gradients, deformation, and microstructural evolution in situ and instantaneously. This opens up opportunities for the integration of physics-based machine learning and ICME approaches to enhance predictive accuracy, adaptability, and decision-making in dynamic production environments.

Workforce development is a pressing concern beyond the aforementioned technical challenges amidst the interdisciplinarity of MM. The interdisciplinary nature of MM spans materials science, robotics, AI, and computational engineering, which requires a new generation of professionals with integrated expertise. Expanding training programs, hands-on prototyping environments, and academia-industry partnerships will be vital to prepare the workforce to innovate and scale MM technologies effectively in ways to ensure return on investment.

In sum, HM and MM show a broad trajectory of digital manufacturing. These approaches go beyond traditional CAD-based and subtractive processes toward intelligent, adaptive systems. Although HM integrates additive and subtractive methods for precision and flexibility, MM introduces a paradigm of real-time property control without fixed tooling. Both technologies align with Industry 4.0 goals of automation and data-driven integration in ways that support Industry 5.0 priorities of sustainability, resilience, and human-centric design. The successful adoption will depend on predictive modeling, sensor integration, robust frameworks, cross-sector collaboration, and investment in skills development, thus, ultimately redefining modern manufacturing as intelligent, sustainable, and adaptive.

References

[1] Raabe D. The materials science behind sustainable metals and alloys, Chem Rev. 2023;123:2436–2608. https://doi.org/10.1021/acs.chemrev.2c00799.

[2] African Union. Agenda 2063: The Africa we want – Background note. 2015;1–20.

[3] Raabe D, Tasan CC, Olivetti EA. Strategies for improving the sustainability of structural metals, Nature. 2019;575:64–74.

[4] Han L, Mu W, Wei S, Liaw PK, Raabe D. Sustainable high-entropy materials?, Sci Adv. 2024;10: eads3926. doi:https://doi.org/10.1126/sciadv.ads3926.

[5] Raabe D, Ponge D, Uggowitzer PJ, Roscher M, Paolantonio M, Liu C. et al. Making sustainable aluminum by recycling scrap: The science of "dirty" alloys, Prog Mater Sci. 2022;128:100947. doi: https://doi.org/10.1016/j.pmatsci.2022.100947.

[6] Hagelüken C, Goldmann D. Recycling and circular economy—towards a closed loop for metals in emerging clean technologies, Miner Econ. 2022;35:539–562. doi:https://doi.org/10.1007/s13563-022-00319-1.

[7] Al Rashid A, Koç M. Additive manufacturing for sustainability and circular economy: Needs, challenges, and opportunities for 3D printing of recycled polymeric waste, Mater Today Sustain. 2023;24:100529. doi:https://doi.org/10.1016/j.mtsust.2023.100529.

[8] Davis C, Hall R, Hazra S, Debattista K, Zhuang S, Duan J. et al. Reuse, remanufacturing and recycling in the steel sector, Philos Trans R Soc A Math Phys Eng Sci. 2024;382:20230244. doi:https://doi.org/10.1098/rsta.2023.0244.

[9] Olivetti EA, Cullen JM. Toward a sustainable materials system, Science. 2018;360:1396–1398. doi: https://doi.org/10.1126/science.aat6821.

[10] Hegab H, Khanna N, Monib N, Salem A. Design for sustainable additive manufacturing: a review, Sustain Mater Technol. 2023;35:e00576: doi:https://doi.org/10.1016/j.susmat.2023.e00576.

[11] Daehn GS, Taub A. Metamorphic manufacturing: The third wave in digital manufacturing, Manuf Lett. 2018;15:86–88. doi:https://doi.org/10.1016/j.mfglet.2018.02.014.

[12] Balasubramanian S, Brennan RW, Norrie DH. An architecture for metamorphic control of holonic manufacturing systems, Comput Ind. 2001;46:13–31.

[13] Szadkowski J. Simplicity concept in structural optimization for the metamorphic manufacturing system, IFAC Proc. 1994;27:431–435. https://doi.org/10.1016/s1474-6670(17)46062-9.

[14] Schweidtmann AM, Zhang D, Von Stosch M. A review and perspective on hybrid modeling methodologies, Digit Chem Eng. 2024;10:100136. doi:https://doi.org/10.1016/j.dche.2023.100136.

[15] Ma S, Ding W, Liu Y, Zhang Y, Ren S, Kong X. et al. Industry 4.0 and cleaner production: A comprehensive review of sustainable and intelligent manufacturing for energy-intensive manufacturing industries, J Clean Prod. 2024;467:142879. doi:https://doi.org/10.1016/j.jclepro.2024.142879.

[16] Dilberoglu UM, Gharehpapagh B, Yaman U, Dolen M. Current trends and research opportunities in hybrid additive manufacturing, Int J Adv Manuf Technol. 2021;113:623–648. doi:https://doi.org/10.1007/s00170-021-06688-1.

[17] Dávila JL, Neto PI, Noritomi PY, Coelho RT, Da Silva JVL. Hybrid manufacturing: a review of the synergy between directed energy deposition and subtractive processes, Int J Adv Manuf Technol. 2020;110:3377–3390. https://doi.org/10.1007/s00170-020-06062-7.

[18] The Minerals Metals and Materials Society. Metamorphic manufacturing: Shaping the future of on-demand components. 2019.

[19] Webster S, Lin H, Carter FM, Ehmann K, Cao J. Physical mechanisms in hybrid additive manufacturing: a process design framework, J Mater Process Technol. 2021;291:117048. doi:https://doi.org/10.1016/j.jmatprotec.2021.117048.

[20] Soshi M, Ring J, Young C, Oda Y, Mori M. Innovative grid molding and cooling using an additive and subtractive hybrid CNC machine tool, CIRP Annals. 2017;66(1):401–404. doi:https://doi.org/10.1016/j.cirp.2017.04.093.
[21] Huang Y, Zeng X. Study on modes of energy action in laser-induction hybrid cladding, Appl Surf Sci. 2009;256:749–756. doi:https://doi.org/10.1016/j.apsusc.2009.08.053.
[22] Strong D, Kay M, Conner B, Wakefield T, Manogharan G. Hybrid manufacturing – integrating traditional manufacturers with additive manufacturing (AM) supply chain, Addit Manuf. 2018;21:159–173. doi:https://doi.org/10.1016/j.addma.2018.03.010.
[23] Manogharan G, Wysk RA, Harrysson OLA. Additive manufacturing-integrated hybrid manufacturing and subtractive processes: economic model and analysis, Int J Comput Integr Manuf. 2016;29:473–488. doi:https://doi.org/10.1080/0951192X.2015.1067920.
[24] Mo X, Hu X, Sun A, Zhang Y. A meta-learning method for smart manufacturing: tool wear prediction using hybrid information under various operating conditions, Robot Comput Integr Manuf. 2025;91:102846. doi:https://doi.org/10.1016/j.rcim.2024.102846.
[25] Ebrahimi Araghizad A, Pashmforoush F, Tehranizadeh F, Kilic K, Budak E. Improving milling force predictions: A hybrid approach integrating physics-based simulation and machine learning for remarkable accuracy across diverse unseen materials and tool types, J Manuf Process. 2024;114:92–107. doi:https://doi.org/10.1016/j.jmapro.2024.02.001.
[26] Kanishka K, Acherjee B. Revolutionizing manufacturing: a comprehensive overview of additive manufacturing processes, materials, developments, and challenges, J Manuf Process. 2023;107:574–619. doi:https://doi.org/10.1016/j.jmapro.2023.10.024.
[27] Tofail SAM, Koumoulos EP, Bandyopadhyay A, Bose S, O'Donoghue L, Charitidis C. Additive manufacturing: Scientific and technological challenges, market uptake and opportunities, Mater Today. 2018;21:22–37.
[28] Bandyopadhyay A, Traxel KD, Lang M, Juhasz M, Eliaz N, Bose S. Alloy design via additive manufacturing: Advantages, challenges, applications and perspectives, Mater Today. 2022;52:207–224. doi:https://doi.org/10.1016/j.mattod.2021.11.026.
[29] Peng T, Kellens K, Tang R, Chen C, Chen G. Sustainability of additive manufacturing: An overview on its energy demand and environmental impact, Addit Manuf. 2018;21:694–704. doi:https://doi.org/10.1016/j.addma.2018.04.022.
[30] Herzog D, Seyda V, Wycisk E, Emmelmann C. Additive manufacturing of metals, Acta Mater. 2016;117:371–392.
[31] Bose S, Ke D, Sahasrabudhe H, Bandyopadhyay A. Additive manufacturing of biomaterials, Prog Mater Sci. 2018;93:45–111. doi:https://doi.org/10.1016/j.pmatsci.2017.08.003.
[32] Zhu D, Feng X, Xu X, Yang Z, Li W, Yan S. et al. Robotic grinding of complex components: A step towards efficient and intelligent machining – challenges, solutions, and applications, Robot Comput Integr Manuf. 2020;65:101908. doi:https://doi.org/10.1016/j.rcim.2019.101908.
[33] Shaik AM, Vvsk R, Rao CS. Development of modular manufacturing systems: A review, Int J Adv Manuf Technol. 2015;76:789–802. https://doi.org/10.1007/s00170-014-6289-2.
[34] Dogra A, Mahna S, Padhee SS, Singla E. Unified modeling of unconventional modular and reconfigurable manipulation system, Robot Comput Integr Manuf. 2022;78:102385. doi:https://doi.org/10.1016/j.rcim.2022.102385.
[35] Lin L, Kollipara PS, Zheng Y. Digital manufacturing of advanced materials: Challenges and perspective, Mater Today. 2019;28:49–62. doi:https://doi.org/10.1016/j.mattod.2019.05.022.
[36] Ingarao G, Priarone PC. A comparative assessment of energy demand and life cycle costs for additive- and subtractive-based manufacturing approaches, J Manuf Process. 2020;56:1219–1229. doi:https://doi.org/10.1016/j.jmapro.2020.06.009.

[37] International Organization for Standardization. Additive manufacturing—General principles—Terminology. ISO/ASTM 52900. 2018. Available from: https://www.iso.org/obp/ui/#iso:std:iso-astm:52900:dis:ed-2:v1:en

[38] Frazier WE. Metal additive manufacturing: A review, J Mater Eng Perform. 2014;23:1917–1928.

[39] Haghnegahdar L, Joshi SS, Dahotre NB. From IoT-based cloud manufacturing approach to intelligent additive manufacturing: Industrial Internet of Things—an overview, Int J Adv Manuf Technol. 2022;119:1461–1478.

[40] Zhou X, Ma J, Zhou W, Welo T. Forming-based geometric correction methods for thin- walled metallic components: a selective review, Int J Adv Manuf Technol. 2023;128:17–39. doi:https://doi.org/10.1007/s00170-023-11948-3.

[41] Wu D, Rosen DW, Wang L, Schaefer D. Cloud-based design and manufacturing: A new paradigm in digital manufacturing and design innovation, Comput Aided Des. 2015;59:1–14. doi:https://doi.org/10.1016/j.cad.2014.07.006.

[42] Chen D, Heyer S, Ibbotson S, Salonitis K, Steingrímsson JG, Thiede S. Direct digital manufacturing: Definition, evolution, and sustainability implications, J Clean Prod. 2015;107:615–625. doi:https://doi.org/10.1016/j.jclepro.2015.05.009.

[43] Chryssolouris G, Mavrikios D, Papakostas N, Mourtzis D, Michalos G, Georgoulias K. Digital manufacturing: history, perspectives, and outlook, Proc Inst Mech Eng B J Eng Manuf. 2009;223:451–462. doi:https://doi.org/10.1243/09544054JEM1241.

[44] Paritala PK, Manchikatla S, Yarlagadda PKDV. Digital manufacturing—applications past, current, and future trends, Procedia Eng. 2017;174:982–991. doi:https://doi.org/10.1016/j.proeng.2017.01.250.

[45] Monostor L, Csáji BC, Kádár B, Pfeiffer A, Ilie-Zudor E, Kemény Z. et al. Towards adaptive and digital manufacturing, Annu Rev Control. 2010;34:118–128. doi:https://doi.org/10.1016/j.arcontrol.2010.02.007.

[46] Bilal DK, Unel M, Tunc LT, Gonul B. Development of a vision-based pose estimation system for robotic machining and improving its accuracy using LSTM neural networks and sparse regression, Robot Comput Integr Manuf. 2022;74:102262. doi:https://doi.org/10.1016/j.rcim.2021.102262.

[47] Xu J, Wang K, Sheng H, Gao M, Zhang S, Tan J. Energy efficiency optimization for ecological 3D printing based on adaptive multi-layer customization, J Clean Prod. 2020;245:118826. doi:https://doi.org/10.1016/j.jclepro.2019.118826.

[48] Wu K, Krewet C, Kuhlenkötter B. Dynamic performance of industrial robot in corner path with CNC controller, Robot Comput Integr Manuf. 2018;54:156–161. doi:https://doi.org/10.1016/j.rcim.2017.11.008.

[49] Misra A, Sharma A, Singh G, Kumar A, Rastogi V. Design and development of a low- cost CNC alternative SCARA robotic arm, Procedia Comput Sci. 2020;171:2459–2468.

[50] Breaz RE, Bologa O, Racz SG. Selecting between CNC milling, robot milling and DMLS processes using a combined AHP and fuzzy approach, Procedia Comput Sci. 2017;122:796–803. doi:https://doi.org/10.1016/j.procs.2017.11.439.

[51] Zhang M, Tao F, Zuo Y, Xiang F, Wang L, Nee AYC. Top ten intelligent algorithms towards smart manufacturing, J Manuf Syst. 2023;71:158–171. doi:10.1016/j.jmsy.2023.09.006.

[52] Park YS, Egilmez G, Kucukvar M. A novel life cycle-based principal component analysis framework for eco-efficiency analysis: Case of the United States manufacturing and transportation nexus, J Clean Prod. 2015;92:327–342. doi:10.1016/j.jclepro.2014.12.057.

[53] Wang P, Li W, Kara S. Dynamic life cycle quantification of metallic elements and their circularity, efficiency, and leakages, J Clean Prod. 2018;174:1492–1502. doi:10.1016/j.jclepro.2017.11.032.

[54] Výtisk J, Honus S, Kočí V, Pagáč M, Hajnyš J, Vujanovic M. et al. Comparative study by life cycle assessment of an air ejector and orifice plate for experimental measuring stand manufactured by conventional manufacturing and additive manufacturing, Sustain Mater Technol. 2022;32:e00431. doi:10.1016/j.susmat.2022.e00431.

[55] Soori M, Ghaleh Jough FK, Dastres R, Arezoo B. Sustainable CNC machining operations: A review, Sustain Oper Comput. 2024;5:73–87. doi:10.1016/j.susoc.2024.01.001.
[56] Shin YC, Chin H, Brink MJ. Characterization of CNC machining centers, J Manuf Syst. 1991;10:407–421.
[57] Lasemi A, Xue D, Gu P. Recent development in CNC machining of freeform surfaces: A state-of-the-art review, Comput Aided Des. 2010;42:641–654. doi:10.1016/j.cad.2010.04.002.
[58] Sparham M, Sarhan AAD, Mardi NA, Hamdi M. Designing and manufacturing an automated lubrication control system in CNC machine tool guideways for more precise machining and less oil consumption, Int J Adv Manuf Technol. 2014;70:1081–1090. doi:10.1007/s00170-013-5325-y.
[59] Kasik DJ, Buxton W, Ferguson DR. Ten CAD challenges, IEEE Comput Graph Appl. 2005;25(2):81–92.
[60] Kasik DJ, Dill JC, Johnson C, Whitton MC. Interactive graphics in industry: The early days, IEEE Comput Graph Appl. 2020;40:89–99. doi:10.1109/MCG.2020.3012228.
[61] Kim J, Pratt MJ, Iyer RG, Sriram RD. Standardized data exchange of CAD models with design intent, Comput Aided Des. 2008;40:760–777. doi:10.1016/j.cad.2007.06.014.
[62] Bettig B, Shah J. Derivation of a standard set of geometric constraints for parametric modeling and data exchange, Comput Aided Des. 2001;33:17–33.
[63] Pratt MJ. Extension of the STEP standard for parametric CAD models, J Comput Inf Sci Eng. 2001;1:269–275. doi:10.1115/1.1419196.
[64] Demeter K, Losonci D, Nagy J. Road to digital manufacturing: a longitudinal case- based analysis, J Manuf Technol Manag. 2021;32:820–839. doi:10.1108/JMTM-06-2019-0226.
[65] Milisavljevic-Syed J, Thames JL, Schaefer D. The digitization of design and manufacturing: A state-of-the-art report on the transition from strategic vision to implementation in industry, Procedia CIRP. 2020;93:575–580. doi:10.1016/j.procir.2020.03.088.
[66] Lu BH. Smart technology renders high-end manufacturing possible, Adv Manuf. 2013;1:107–111. doi:10.1007/s40436-013-0028-z.
[67] Bankapalli NK, Gupta V, Saxena P, Bajpai A, Lahoda C, Polte J. Filament fabrication and subsequent additive manufacturing, debinding, and sintering for extrusion-based metal additive manufacturing and their applications: A review, Compos B Eng. 2023;264:110915. doi:10.1016/j.compositesb.2023.110915.
[68] Braniff A, Akundi SS, Liu Y, Dantas B, Niknezhad SS, Khan F. et al. Real-time process safety and systems decision-making toward safe and smart chemical manufacturing, Digit Chem Eng. 2025;15:100227. doi:10.1016/j.dche.2025.100227.
[69] Soleymanizadeh H, Qu Q, Bamakan SMH, Zanjirchi S. Digital twin empowering manufacturing paradigms: Lean, agile, just-in-time (JIT), flexible, resilience, sustainable, Procedia Comput Sci. 2023;227:1258–1267. doi:10.1016/j.procs.2023.08.114.
[70] Tao F, Qi Q. Make more digital twins, Nature. 2019;573:490–491.
[71] Kumar R, Agrawal N. Shaping the future of industry: understanding the dynamics of industrial digital twins, Comput Ind Eng. 2024;191:110172. doi:10.1016/j.cie.2024.110172.
[72] Zafar MH, Langås EF, Sanfilippo F. Exploring the synergies between collaborative robotics, digital twins, augmentation, and Industry 5.0 for smart manufacturing: A state- of-the-art review, Robot Comput Integr Manuf. 2024;89:102769. doi:10.1016/j.rcim.2024.102769.
[73] Helo P, Hao Y, Toshev R, Boldosova V. Cloud manufacturing ecosystem analysis and design, Robot Comput Integr Manuf. 2021;67:102050. doi:10.1016/j.rcim.2020.102050.
[74] Simeone A, Caggiano A, Boun L, Deng B. Intelligent cloud manufacturing platform for efficient resource sharing in smart manufacturing networks, Procedia CIRP. 2019;81:233–238. doi:10.1016/j.procir.2019.02.056.
[75] García-Collado A, Blanco JM, Gupta MK, Dorado-Vicente R. Advances in polymers- based multi-material additive manufacturing techniques: State-of-art review on properties and applications, Addit Manuf. 2022;50:102577. doi:10.1016/j.addma.2021.102577.

[76] Wang C, Tan XP, Tor SB, Lim CS. Machine learning in additive manufacturing: state- of-the-art and perspectives, Addit Manuf. 2020;36:101538. doi:10.1016/j.addma.2020.101538.
[77] Bajaj P, Hariharan A, Kini A, Kürnsteiner P, Raabe D, Jägle EA. Steels in additive manufacturing: A review of their microstructure and properties, Mater Sci Eng A. 2020;772:138633.
[78] Xie Y, Zhang H, Zhou F. Improvement in geometrical accuracy and mechanical property for arc-based additive manufacturing using metamorphic rolling mechanism, J Manuf Sci Eng. 2016;138:1–9. doi:10.1115/1.4032079.
[79] Hjaila K, Laínez-Aguirre JM, Puigjaner L, Espuña A. Decentralized manufacturing supply chains coordination under uncertain competitiveness, Procedia Eng. 2015;132:942–949. doi:10.1016/j.proeng.2015.12.581.
[80] Mourtzis D, Doukas M. A web-based platform for customer integration in the decentralised manufacturing of personalised products, Procedia CIRP. 2012;3:209–214. doi:10.1016/j.procir.2012.07.037.
[81] Bochmann L, Bänziger T, Kunz A, Wegener K. Human-robot collaboration in decentralized manufacturing systems: An approach for simulation-based evaluation of future intelligent production, Procedia CIRP. 2017;63:624–629. doi:10.1016/j.procir.2016.06.021.
[82] Xiong Y, Tang Y, Kim S, Rosen DW. Human-machine collaborative additive manufacturing, J Manuf Syst. 2023;66:82–91. doi:10.1016/j.jmsy.2022.12.004.
[83] Yoshikawa H. Manufacturing and the 21st century—intelligent manufacturing systems and the renaissance of the manufacturing industry, Technol Forecast Soc Change. 1995;49:195–213.
[84] Bagger C, Olsen FO. Review of laser hybrid welding, J Laser Appl. 2005;17:2–14. doi:10.2351/1.1848532.
[85] Zhou T, Gani R, Sundmacher K. Hybrid data-driven and mechanistic modeling approaches for multiscale material and process design, Engineering. 2021;7:1231–1238. doi:10.1016/j.eng.2020.12.022.
[86] Hernández-Nava E, Mahoney P, Smith CJ, Donoghue J, Todd I, Tammas-Williams S. Additive manufacturing titanium components with isotropic or graded properties by hybrid electron beam melting/hot isostatic pressing powder processing, Sci Rep. 2019;9:1–11.
[87] Klenam DEP, Bamisaye OS, Williams IE, van der Merwe JW, Bodunrin MO. Global perspective and African outlook on additive manufacturing research: An overview, Manuf Rev (Les Ulis). 2022;9:1–38. doi:10.1051/mfreview/2022033.
[88] Adekanye SA, Mahamood RM, Akinlabi ET, Owolabi MG. Additive manufacturing: the future of manufacturing, Mater Technol. 2017;51:709–715.
[89] Cooke S, Ahmadi K, Willerth S, Herring R. Metal additive manufacturing: technology, metallurgy and modelling, J Manuf Process. 2020;57:978–1003.
[90] Keller AZ, Kamath ARR, Perera UD. Reliability analysis of CNC machine tools, Reliab Eng. 1982;3:449–473.
[91] Vichare P, Nassehi A, Kumar S, Newman ST. A unified manufacturing resource model for representing CNC machining systems, Robot Comput Integr Manuf. 2009;25(999):1007. doi:10.1016/j.rcim.2009.04.014.
[92] Henci H. An overview of proses CNC machining, J Mech Sci Eng. 2019;6:19–23.
[93] Putz M, Tagliaferri F, Dix M, Wertheim R. Methodology for analysing hybrid cutting processes with thermal assistance, Prod Eng. 2015;9:537–549. doi:10.1007/s11740-015-0624-9.
[94] Lauwers B, Klocke F, Klink A, Tekkaya AE, Neugebauer R, McIntosh D. Hybrid processes in manufacturing, CIRP Ann Manuf Technol. 2014;63:561–583. doi:10.1016/j.cirp.2014.05.003.
[95] Holtewert P, Wutzke R, Seidelmann J, Bauernhansl T. Virtual Fort Knox: Federative, secure and cloud-based platform for manufacturing, Procedia CIRP. 2013;7:527–532. doi:10.1016/j.procir.2013.06.027.

[96] Li C, Jiang Y, Fang P. Innovation in advanced manufacturing driven by supercomputing, Procedia CIRP. 2019;81:584–589. doi:10.1016/j.procir.2019.04.092.
[97] Salvi H, Vesuwala H, Raval P, Badheka V, Khanna N. Sustainability analysis of additive + subtractive manufacturing processes for Inconel 625, Sustain Mater Technol. 2023;35:e00580: doi:10.1016/j.susmat.2023.e00580.
[98] Feldhausen T, Heinrich L, Saleeby K, Burl A, Post B, MacDonald E. et al. Review of computer-aided manufacturing (CAM) strategies for hybrid directed energy deposition, Addit Manuf. 2022;56:102900. doi:10.1016/j.addma.2022.102900.
[99] Jiménez A, Bidare P, Hassanin H, Tarlochan F, Dimov S, Essa K. Powder-based laser hybrid additive manufacturing of metals: a review, Int J Adv Manuf Technol. 2021;114:63–93. doi:10.1007/s00170-021-06855-4.
[100] Liu G, Lu X, Zhang X, Zhao Y, Yi S, Xu J. et al. 3D/4D additive–subtractive manufacturing of heterogeneous ceramics, J Mater Sci Technol. 2024;201:210–221. doi:10.1016/j.jmst.2024.02.069.
[101] Zhang H, Liu J, Zhu X, Zhang Y, Li H, Li Z. et al. Hybrid additive-subtractive manufacturing of high-density copper-based flexible transparent circuits based on side- etch process, J Manuf Process. 2024;125:580–588. doi:10.1016/j.jmapro.2024.07.098.
[102] Qian M, Song S, Yang Z, Guo Y, Wang K. A comparison of wire-feed additive manufacturing and hybrid additive/subtractive manufacturing of high nitrogen steel, J Mater Res Technol. 2024;33:4092–4104. doi:10.1016/j.jmrt.2024.10.090.
[103] Wang J, Wang H, Gao H, Cheng F, Zhang B, Zhang M. et al. Microstructure characteristics and failure mechanisms of hybrid manufacturing of FGH96 and IC10 bimetal component using laser directed energy deposition, Mater Sci Eng A. 2024;889:145965. doi:10.1016/j.msea.2023.145965.
[104] Sampaio RFV, Pragana JPM, Bragança IMF, Silva CMA, Nielsen CV, Martins PAF. Modelling of wire-arc additive manufacturing – a review, Adv Ind Manuf Eng. 2023;6:100121. doi:10.1016/j.aime.2023.100121.
[105] Li K, Chen W, Gong N, Pu H, Luo J, Zhang DZ. et al. A critical review on wire-arc directed energy deposition of high-performance steels, J Mater Res Technol. 2023;24:9369–9412. doi:10.1016/j.jmrt.2023.05.163.
[106] Derekar KS. A review of wire arc additive manufacturing and advances in wire arc additive manufacturing of aluminium, Mater Sci Technol (UK). 2018;34:895–916.
[107] Li H, Shi X, Wu B, Corradi DR, Pan Z, Li H. Wire arc additive manufacturing: A review on digital twinning and visualization process, J Manuf Process. 2024;116(293):305. doi:10.1016/j.jmapro.2024.03.001.
[108] Xia C, Pan Z, Polden J, Li H, Xu Y, Chen S. et al. A review on wire arc additive manufacturing: Monitoring, control and a framework of automated system, J Manuf Syst. 2020;57:31–45. doi:10.1016/j.jmsy.2020.08.008.
[109] Shen Q, Kong X, Chen X. Fabrication of bulk Al–Co–Cr–Fe–Ni high-entropy alloy using combined cable wire arc additive manufacturing (CCW-AAM): Microstructure and mechanical properties, J Mater Sci Technol. 2021;74:136–142.
[110] Duan X, Li Q, Xie W, Yang X. Wire arc metal additive manufacturing using pulsed arc plasma (PAP-WAAM) for effective heat management, J Mater Process Technol. 2023;311:117806. doi:10.1016/j.jmatprotec.2022.117806.
[111] Duan X, Yang X. Achieving enhanced strength and ductility in 316L stainless steel via wire arc additive manufacturing using pulsed arc plasma, Mater Sci Eng A. 2023;867:144711. doi:10.1016/j.msea.2023.144711.
[112] Jia C, Liu W, Chen M, Guo M, Wu S, Wu C. Investigation on arc plasma, droplet, and molten pool behaviours in compulsively constricted WAAM, Addit Manuf. 2020;34:101235. doi:10.1016/j.addma.2020.101235.

[113] Klenam D, Asumadu T, Bodunrin M, Vandadi M, Bond T, Van der Merwe J. et al. Cold spray coatings of complex concentrated alloys: Critical assessment of milestones, challenges, and opportunities, Coatings. 2023;13:1–50.
[114] Li D. A review of microstructure evolution during ultrasonic additive manufacturing, Int J Adv Manuf Technol. 2021;113:1–19. doi:10.1007/s00170-020-06439-8.
[115] Poudel L, Jha S, Meeker R, Phan DN, Bhowmik R. Advanced predictive quality assessment for ultrasonic additive manufacturing with deep learning model, J Intell Manuf. 2025; doi:10.1007/s10845-025-02582-9.
[116] Al-Naggar YM, Jamil N, Hassan MF, Yusoff AR. Condition monitoring based on IoT for predictive maintenance of CNC machines, Procedia CIRP. 2021;102:314–318. doi:10.1016/j.procir.2021.09.054.
[117] Byiringiro JB, Kim MY, Ko TJ. Process modeling of hybrid machining system consisted of electro discharge machining and end milling, Int J Adv Manuf Technol. 2012;60:1247–1254. doi:10.1007/s00170-012-4089-0.
[118] Mognol P, Jégou L, Rivette M, Furet B. High speed milling, electro discharge machining and direct metal laser sintering: a method to optimize these processes in hybrid rapid tooling, Int J Adv Manuf Technol. 2006;29:35–40. 10.1007/s00170-005-2502-7
[119] Zolfagharian A, Bodaghi M, Le DA. Editorial: 4D printing and 3D printing in robotics, sensors, and actuators manufacturing, Front Robot AI. 2022;9:1110571. doi:10.3389/frobt.2022.1110571.
[120] Liu Y, Wang L, Makris S, Krüger J. Smart robotics for manufacturing, Robot Comput Integr Manuf. 2023;102535: doi:10.1016/j.rcim.2023.102535.
[121] Unune DR, Mali HS. Current status and applications of hybrid micro-machining processes: A review, Proc Inst Mech Eng B J Eng Manuf. 2015;229:1681–1693. doi:10.1177/0954405414546141.
[122] Schweiger J, Edelhoff D, Güth JF. 3D printing in digital prosthetic dentistry: An overview of recent developments in additive manufacturing, J Clin Med. 2021;10:2010. doi:10.3390/jcm10092010.
[123] Ibrahim P, Garrard R, Penchev P, Man K, Cox SC, Dimov S. et al. Hybrid manufacturing and performance evaluation of β Ti-alloy stents, Mater Des. 2024;247:113420. doi:10.1016/j.matdes.2024.113420.
[124] Vardhan KH, Kumar PS, Panda RC. A review on heavy metal pollution, toxicity and remedial measures: current trends and future perspectives, J Mol Liq. 2019;290:111197. doi:10.1016/j.molliq.2019.111197.
[125] Blakey-Milner B, Gradl P, Snedden G, Brooks M, Pitot J, Lopez E. et al. Metal additive manufacturing in aerospace: A review, Mater Des. 2021;209:110008.
[126] Ghorbani H, Khameneifar F. Construction of damage-free digital twin of damaged aero-engine blades for repair volume generation in remanufacturing, Robot Comput Integr Manuf. 2022;77:102335. doi:10.1016/j.rcim.2022.102335.
[127] Xiao W, Liu G, Zhao G. Generating the tool path directly with point cloud for aero- engine blades repair, Proc Inst Mech Eng B J Eng Manuf. 2021;235:877–886. doi:10.1177/0954405420970915.
[128] Wahlman A, Brian C, Drinkwine M, Army US. The M1 Abrams today and tomorrow, Mil Rev. 2014;11–20.
[129] Dogan E, Bhusal A, Cecen B, Miri AK. 3D printing metamaterials towards tissue engineering, Appl Mater Today. 2020;20:100752.
[130] Bose S, Vahabzadeh S, Bandyopadhyay A. Bone tissue engineering using 3D printing, Mater Today. 2013;16:496–504.
[131] MacDonald E, Wicker R. Multiprocess 3D printing for increasing component functionality, Science. 2016;353:aaf2093-1–aaf2093-8.
[132] Colpani A, Fiorentino A, Ceretti E. 3D printing for health and wealth: Fabrication of custom-made medical devices through additive manufacturing, AIP Conf Proc. 2018; doi:10.1063/1.5034998.

[133] Suriano JT, Tafuni A, Mullen L, Racanelli J, Tarantino R, Lieber SC. Medical device hybrid manufacturing: Translating the coordinate system from metal additive manufacturing to subtractive post-processing, ASME Open J Eng. 2023;2: doi:10.1115/1.4062187.

[134] Ahn SH, Chun DM, Kim CS. Nanoscale hybrid manufacturing process by nanoparticle deposition system (NPDS) and focused ion beam (FIB), CIRP Ann Manuf Technol. 2011;60:583–586. doi:10.1016/j.cirp.2011.03.071.

[135] Chalicheemalapalli Jayasankar D, Gnaase S, Kaiser MA, Lehnert D, Tröster T. Advancements in hybrid additive manufacturing: Integrating SLM and LMD for high- performance applications, Metals (Basel). 2024;14: doi:10.3390/met14070772.

[136] Krimpenis AA, Noeas GD. Application of hybrid manufacturing processes in microfabrication, J Manuf Process. 2022;80:328–346. doi:10.1016/j.jmapro.2022.06.009.

[137] Helfman R, John M, Hanratty DMT TED–Turbine Engine Diagnostics: An expert diagnostic system for the M1 Abrams AGT1500 turbine engine, 1995.

[138] The Minerals, Metals & Materials Society (TMS). Employing artificial intelligence to accelerate development and implementation of materials and manufacturing innovations. Pittsburgh: The Minerals, Metals & Materials Society; 2022.

[139] Camarinha-Matos LM, Rocha AD, Graça P. Collaborative approaches in sustainable and resilient manufacturing, J Intell Manuf. 2024;35:499–519. 10.1007/s10845-022-02060-6

[140] Majeed A, Zhang Y, Ren S, Lv J, Peng T, Waqar S, Yin E. A big data-driven framework for sustainable and smart additive manufacturing, Robot Comput Integr Manuf. 2021;67:102026.

[141] Gault R, Pham DT. A comparison of rapid prototyping technologies, Int J Mach Tools Manuf. 1998;38:1257–1287. doi:10.1016/S0890-6955(97)00137-5.

[142] Gibson I, Rosen D, Brent S. Additive Manufacturing Technologies: 3D Printing, Rapid Prototyping, and Direct Digital Manufacturing. 2nd ed, New York: Springer Science+Business Media; 2015.

[143] Dimitrov D, Schreve K, Taylor A, Vincent B. Rapid prototyping driven design and realisation of large components, Rapid Prototyp J. 2007;13:85–91.

[144] Springer H, Raabe D. Rapid alloy prototyping: Compositional and thermo-mechanical high throughput bulk combinatorial design of structural materials based on the example of 30Mn-1.2C-xAl triplex steels, Acta Mater. 2012;60:4950–4959.

[145] Ali A, Hu C, Yttri EA, Panat R. Recent advances in 3D printing of biomedical sensing devices, Adv Funct Mater. 2022;32:1–22.

[146] Souri H, Banerjee H, Jusufi A, Radacsi N, Stokes AA, Park I, Sitti M, Amjadi M. Wearable and stretchable strain sensors: Materials, sensing mechanisms, and applications, Adv Intell Syst. 2020;2:2000039. doi:10.1002/aisy.202000039.

[147] Rescsanski S, Hebert R, Haghighi A, Tang J, Imani F. Towards intelligent cooperative robotics in additive manufacturing: Past, present, and future, Robot Comput Integr Manuf. 2025;93:102925. doi:10.1016/j.rcim.2024.102925.

Rakesh Kumar, Ravi Pratap Singh, Sandeep Singhal, Pradeep Kumar, Mustafizur Rahman

2 Exploring the Machinability of Macor Dental Ceramic via Electrochemical Discharge Machining Process

Abstract: Micromachining of ceramic materials has gained significant attention in biomedical applications, particularly in dental restorations where Macor ceramic materials are widely used. Electrochemical discharge machining (ECDM) is preferred for micromanufacturing of complex micro holes, particularly in materials like Macor dental ceramic, which are widely used in medical and lab equipment, aerospace, microelectronics, automobile industries, etc. The present study presents an investigation on the stimulus of various input parameters on machining features in ECDM of Macor dental ceramic. The impact of key variables such as electrolyte concentration, applied voltage, and feed rate has been examined on process responses of interest, i.e., cutting rate (CR) and depth of cut of machined holes to make the process more sustainable and precision machining of hard materials in microfabrication applications. The experimental results demonstrated a maximum CR of 233.5 µg/min, achieved at an applied voltage of 72 V, an electrolyte concentration of 32 g/L, and a feed rate of 0.12 mm/min. Conversely, the deepest microhole (1,410 µm) was obtained under slightly different conditions: 72 V, 28 g/L electrolyte concentration, and a reduced feed rate of 0.10 mm/min, suggesting a trade-off between machining speed and penetration depth. Furthermore, comprehensive surface characterization of the fabricated microholes was conducted using scanning electron microscopy, energy-dispersive X-ray spectroscopy, and optical microscopy to evaluate morphological features, dimensional accuracy, and potential recast layer formation.

Keywords: ECDM, microholes, depth of cut, cutting rate, Macor ceramic

2.1 Introduction

Macor, a machinable glass ceramic, belongs to the class of advanced engineering ceramics that exhibit an exceptional balance of thermal stability, mechanical strength, electrical insulation, and chemical inertness. These inherent mechanical and chemical characteristics, including exceptional hardness, superior wear resistance, and outstanding corrosion resistance, and the ability to retain dimensional stability at elevated temperatures, make Macor a suitable candidate for a diverse array of high-performance applications [1, 2]. Its widespread use spans across critical sectors including aerospace, electronics, medical instrumentation, and analytical equipment. In particular, Macor

 | https://doi.org/10.1515/9783111704456-002

has gained significant attention in dental and biomedical domains due to its biocompatibility, low thermal conductivity, and structural precision, making it ideal for dental prosthetics and lab-on-chip platforms [3, 4]. Despite these advantages, the brittle nature and high hardness of Macor pose significant challenges during machining. Conventional machining techniques, including turning, drilling, and grinding, often result in surface damage, dimensional inaccuracies, low material removal rates (MRRs), and increased tool wear (TW), thereby limiting their effectiveness in producing precision microfeatures required for dental applications. To address these constraints, advanced nonconventional machining techniques like electrochemical discharge machining (ECDM) have emerged as promising alternatives [5, 6]. ECDM, a hybrid process that combines electrochemical and spark erosion mechanisms, offers a noncontact, thermally assisted material removal technique capable of machining insulating and brittle materials like Macor with improved accuracy and reduced surface defects [7]. This method is particularly advantageous for fabricating intricate and miniaturized features in dental ceramics, which are otherwise hard to achieve using conventional tools.

The ECDM process has arisen as an auspicious hybrid nontraditional machining technique, especially suitable for insulating and brittle materials like Macor machinable glass ceramic. The process utilizes a conductive tool electrode and an auxiliary electrode, both deeper in a chemical solution, commonly a strong alkaline solution, such as NaOH and KOH [8]. When a sufficient DC voltage, typically in the range of 40–70 V, is applied across the electrodes, electrolysis is initiated at the tool-electrolyte interface, leading to the establishment of hydrogen gas bubbles on the tool surface. As the applied voltage surpasses a critical threshold, these gas bubbles unite to form a stable and insulating gas layer around the tool electrode. Upon further increase in voltage, dielectric breakdown of this gas film occurs, resulting in a series of transient spark energy between the tool and the surrounding electrolyte. These discharges generate localized high-temperature zones, which cause the material beneath the tool tip to melt and vaporize, thereby enabling material removal through predominantly thermal mechanisms [9–11]. Although electrochemical reactions play a secondary role in the overall material removal process, they are essential in assisting debris removal and sustaining the discharge environment, contributing to stable machining conditions [12, 13]. The schematic presentation of ECDM of Macor is illustrated in Fig. 2.1, highlighting the stages of gas film formation, spark initiation, and subsequent material removal.

Numerous researchers have explored the applicability of ECDM in machining nonconductive ceramics. For instance, Arya et al. [14] investigated adaptive tool feeding strategies and parameter optimization for conventional ECDM, reporting that a feed rate of 25.15 μm/s, working gap of 8.56 μm, and dwell time of 904.04 ms were optimal for achieving maximum depth and minimal entrance diameter. Similarly, Tiwari and Panda [15] demonstrated that using a nickel-coated copper electrode in glass drilling significantly reduced hole overcut (OC) and the heat-affected zone (HAZ), while improving hole circularity in comparison to an uncoated copper tool. Sandhu

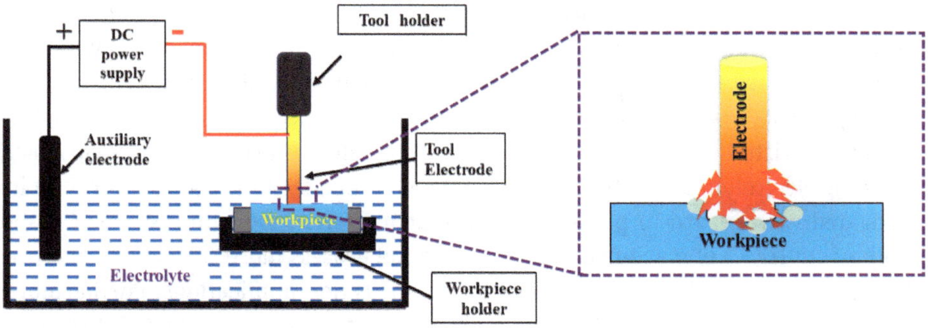

Fig. 2.1: Illustrative schematic of the ECDM process.

et al. [16] performed both single-objective and multiobjective optimization studies to identify the best set of process parameters. The single-objective optimization demonstrated substantial performance gains, exhibiting 120.24% improvement in MRR, 69.14% reduction in OC, and 77.45% decrease in TW. Conversely, the multiobjective approach presented more balanced enhancements, showing a 25.01% increase in MRR and a 22.55% reduction in TW, albeit with a 16.41% compromise in OC. Bellubbi et al. [17] examined the multioptimization methods of ECDM parameters using the Taguchi-GRA technique. It was observed that the optimum condition of parameters was 45 V, 25 wt%, and pulse frequency 400 Hz, and based on the Taguchi-GRA technique, applied voltage was more influencing on MRR, OC, and TWR. Antil [18] implemented an ANN-based hybrid optimization scheme for ECDM parameter optimization when machining silicon carbide reinforced epoxy composites. Their desirability function approach revealed optimal process parameters of 65 V applied voltage, 90.003 g/L electrolyte concentration, 70 mm IEG, and 0.535 duty ratio. Their study also revealed that simultaneous increases in input parameters led to higher cutting rate (CR) (MRR) and a reduction in hole taper. Madhavi and Hiremath [3] conducted experimental research on characterization and fabrication of channels in quartz glass using the micro-ECDM process. They analyzed the process parameters with the GRA technique to determine the optimal settings for improved results. The GRA optimized parameters were found to be −40 V, 20 wt%, and 50% DF resultant in 81 µg/m of MRR, 2.99 µg/m of TWR, and 557.6 µm width of the channel. Bhargav et al. [19] carried out an experimental examination on micromachining of PMMA using µ-ECDM system and study the multioptimization using the JAYA algorithm and MADM R-method and the effect of tool rotation. Multioptimization yields 51 V, 24 wt%, and 55% DF as the ideal process parameters. Kumar and Singh [20] analyzed the influence of auxiliary electrodes and electrolyte properties during microchannel fabrication in Pyrex glass, achieving a MRR of 0.4134 mg/min and an OC width of 40 µm under optimized conditions.

While extensive efforts have been made to enhance ECDM performance for microlevel machining of glass and polymeric substrates, the application of this technique to macroscale nonconductive ceramic materials remains largely unexplored. Most existing studies are confined to micromachining and involve materials with relatively lower hardness and thermal resistance compared to advanced ceramics. This study systematically examines the influence of critical process parameters on machining characteristics during ECDM of Macor dental ceramic. The effects of key variables, including electrolyte concentration, applied voltage, and feed rate, were rigorously investigated with regard to process response characteristics of interest, i.e., CR and depth of cut (DOC) of machined holes, to make the process more sustainable and precision machining of hard materials in microfabrication applications. Furthermore, the surface characteristics of the fabricated microholes were comprehensively characterized using scanning electron microscopy (SEM) for topographical analysis, energy-dispersive X-ray spectroscopy (EDS) for elemental composition, and optical profilometry for dimensional assessment.

2.2 Experimental Methodology

The Macor ceramic material used in this study comprises of approximately 45% borosilicate glass and 55% fluorophlogopite mica. Its unique microstructural composition sets it apart from conventional ceramics and glass materials, offering distinctive properties that are advantageous for precision machining applications. The experimental investigation explores the fabrication of microdrills on Macor ceramic using an in-house developed ECDM apparatus. The experimental setup incorporated a three-axis precision motion control system, pledging micronlevel accuracy and process repeatability throughout machining operations. This specialized configuration was tailored to address the distinct contests of machining Macor dental ceramics, which demand controlled processing conditions owing to their dielectric properties and thermal sensitivity. As illustrated in Fig. 2.2, the system features a three-axis precision movement (*X*, *Y*, *Z*) controlled via Mach3 software, with signal interfacing handled by an Arduino Uno and stepper motor driver. A minimal feed rate of 1 µm/min, which ensures high-precision machining, was implemented to achieve micronlevel machining accuracy. The electrolyte reservoir, fabricated from a 3 mm translucent acrylic sheet, is mounted on the *X*-*Y* axes to confirm exact workpiece alignment. The fixture, machined from a nylon block and secured with PLA-based 3D-printed clamps, was chosen for its nonconductive and chemically inert characteristics. The tool electrode, a 550 µm diameter stainless steel rod, was utilized due to its optimal combination of electrical conductance and wear resistance. The experimental setup employed a regulated DC power supply (0–100 V, 0–10 A) with polarity configured such that the tool electrode served as the cathode, while the auxiliary electrode of stainless steel plate acted as the anode, connected via corrosion-resistant clamps. Discharge characteristics were continuously recorded using a digital oscilloscope to analyze spark stability and energy distribution. The interelectrode gap was minimized (<50 μm) to concentrate discharge energy, facilitating material removal through three concurrent mechanisms: (i) localized melting by spark-induced plasma, (ii) vaporization of molten material, and (iii) electrochemical dissolution. NaOH electrolyte was chosen after a systematic review of prior work, as it demonstrated a higher MRR compared to other electrolyte solutions. During experimentation, the electrolyte level was precisely controlled at 2.0 ± 0.1 mm above the workpiece surface using a graduated reservoir, maintaining consistent machining conditions.

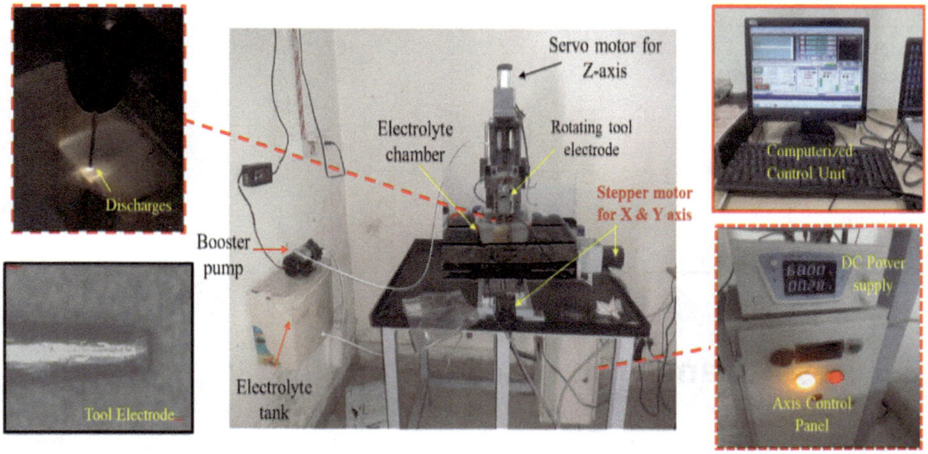

Fig. 2.2: Custom-designed in-house ECDM experimental setup.

The critical process parameters and their operational ranges were established through an extensive review of existing literature, coupled with systematic preliminary testing. The experimental design was formulated using a Box-Behnken design (BBD) approach under the response surface methodology (RSM) framework. This design enabled a systematic investigation of applied voltage, feed rate, and electrolyte concentration effects on performance metrics, specifically CR and microhole depth. The CR was quantitatively evaluated using eq. (1), while dimensional characterization of the machined microholes was performed using a high-precision Mitutoyo dial gauge (Model 2109SB-10) with a measurement accuracy of ±0.005 mm. This measurement instrument has been extensively validated in prior ECDM research, demonstrating consistent measurement reliability [21]. Postmachining analysis included comprehensive surface characterization, utilizing optical microscopy for preliminary feature assessment, followed by advanced SEM for detailed microstructural evaluation. EDS complemented these analyses by providing elemental composition data. This multimodal characterization approach enabled a thorough investigation of thermal-induced defects and surface quality parameters. The experimental program was systematically designed using the Box-Behnken methodology, with all test conditions strictly adhering to the predefined parameter combinations specified in Tab. 2.1:

$$CR = \frac{weight\ of\ sample\ before\ machining - weight\ of\ sample\ after\ machining}{machining\ time} \tag{1}$$

Tab. 2.1: Selected process parameters anc their corresponding levels for experimentation.

Factors/ (unit)	**Levels**		
	−1	**0**	**+1**
Voltage (V)	62	67	72
Electrolyte Concentration (g/L)	24	28	32
Feed rate (mm/min)	0.1	0.12	0.14
Constant parameters			
Electrolyte		NaOH	
IEG (mm)		30	
Tool cross section		Circular	

2.3 Results and Discussion

This study systematically investigates the influence of critical process variables on machining outcomes during microhole generation in Macor dental ceramics via ECDM. Three primary input parameters, applied voltage, electrolyte concentration, and feed rate, were evaluated for their effects on performance metrics, including CR, penetration depth, and surface integrity attributes (particularly HAZ formation). A design of experiments (DOE) framework was adopted, employing RSM with a BBD to enable comprehensive yet efficient parameter space exploration. This approach facilitated rigorous analysis of variable interactions, while minimizing experimental runs. The DOE and statistical evaluation were performed using Design-Expert 13.0 statistical software. Following the establishment of the design matrix, microhole fabrication was carried out on Macor ceramic workpieces across varying process parameter combinations. To ensure data reliability, all experimental trials were replicated twice under identical conditions. The complete experimental design, including the averaged results for CR and DOC, is systematically presented in Tab. 2.2.

Tab. 2.2: Experimental layout with corresponding values of CR and DOC.

Run	AV	EC	FR	CR (μg/min)	DOC (μm)
1	72	32	0.12	233.53	1,390
2	62	28	0.1	115	1,055
3	67	32	0.1	185	1,245
4	67	28	0.12	161.67	1,184
5	72	24	0.12	137.72	1,028
6	67	24	0.1	160	1,234
7	72	28	0.14	209.79	965
8	67	32	0.14	205.73	890
9	67	24	0.14	189.32	735
10	62	24	0.12	179.64	856
11	72	28	0.1	215.58	1,410
12	67	28	0.12	158.37	1,195
13	62	28	0.14	160.83	775
14	62	32	0.12	185.62	1,115
15	67	28	0.12	166.51	1,175

2.3.1 Impact of Machining Variables on Cutting Rate

In the ECDM process, applied voltage emerges as a critical operational parameter that significantly governs machining performance, especially affecting material removal efficiency and penetration depth during microhole generation [22]. The applied voltage directly influences the development of gas bubbles (hydrogen and oxygen) due to electrolysis. Beyond the critical voltage, these bubbles form a dielectric gaseous layer around the tool electrode, leading to thermal spark discharges that erode the ceramic workpiece [23]. At lower applied potential, insufficient discharge energy prevents stable plasma formation, leading to intermittent spark generation and consequently nonuniform material erosion. As voltage increases, the development of gas bubbles becomes more consistent and denser near the tool electrode, which in turn leads to uniform energy dispersion along the electrode periphery and facilitates balanced material erosion through simultaneous thermal ablation and molten-phase expulsion. Furthermore, at excessively high voltages, this results in more intense sparks and higher material removal. However, excessive voltage can cause uncontrolled sparking, wider HAZs, and TW, reducing precision and dimensional accuracy. In the current study, the applied voltage varies from 62 to 72 V, and their corresponding CR values were calculated using eq. (1). Fig. 2.3(a) presents the influences of applied voltage on material CR of the fabricated microholes on Macor dental ceramic. The CR increases from 115 to 233.5 μg/min, as applied voltage rises from 62 to 72 V. The CR is directly proportional to applied voltage, but at higher values of voltage, thermal cracks and nonuniform machined surface is obtained, which affect the dimensional precision of the fabricated microholes.

Further, the electrolyte concentration also plays a substantial role in the ECDM process. Electrolyte concentration affects the ionic conductivity of the medium and plays a dynamic role in bubble formation and discharge initiation. The electrolyte composition significantly influences ECDM performance through two primary mechanisms: ionic transport efficiency and thermal gradient regulation at the electrode interface. Elevated electrolyte concentrations improve solution conductivity by facilitating greater ion mobility, which subsequently intensifies charge transfer rates and discharge frequency. While this enhances material removal efficiency through more vigorous electrochemical activity, it simultaneously introduces challenges including pronounced thermal gradients and hydrodynamic instabilities that may compromise surface finish quality [24]. Experiments conducted with NaOH concentrations ranging from 24 to 32 g/L revealed a positive correlation between concentration and CR, as illustrated in Fig. 2.3(b). On the other hand, when the electrolyte concentration is low, the reduced number of charge carriers weakens the discharge intensity, resulting in lower energy input and limited material removal. This can lead to smoother surfaces in some cases but may also cause inconsistent erosion due to unstable discharge behavior, producing irregular surface patterns [25]. At moderate concentrations, discharge activity becomes more stable and uniform, facilitating consistent material removal and yielding smoother microchannel surfaces with fewer defects. Topographical variations analyzed using an optical microscope in ECDM-generated microholes, corresponding to different parameter combinations, are visualized in Fig. 2.4. From the experimentation, it was analyzed that the CR conversely increases as the electrolyte concentration increased. The CR increased by approximately 40% as electrolyte concentration rises from 24 to 32 g/L.

Furthermore, the feed rate of the tool relative to the workpiece plays a pivotal role in determining the overall machining performance. An increase in feed rate typically results in a higher CR, thereby enhancing machining efficiency. This is primarily because more material is engaged with the tool per unit time, leading to a higher volumetric removal [7]. However, an excessively high feed rate can adversely affect the process. At elevated feed rates, a notable decline in material removal was observed after a certain duration. This reduction is primarily attributed to unstable spark generation around the tool electrode, which arises due to the minimal electrode-workpiece gap caused by the rapid tool advancement. Consequently, as the tool continues to feed with insufficient clearance, it begins to deform or bend, as illustrated in Fig. 2.5. It may compromise the surface integrity of the machined part, leading to a rougher surface finish due to inadequate interaction time between the cathode electrode and the workpiece. Additionally, it can accelerate TW and thermal degradation, thereby reducing tool life and dimensional accuracy. Experimental observations reveal a nearly linear relationship between feed rate and CR up to an optimal threshold, as shown in Fig. 2.3(c). Beyond this point, further increase in feed rate led to a decline in cutting efficiency. This reduction can be attributed to insufficient generation of the gas layer around the tool, which is essential in these processes. The gas layer serves as an insulator barrier and facilitates the generation of spark discharges. At higher

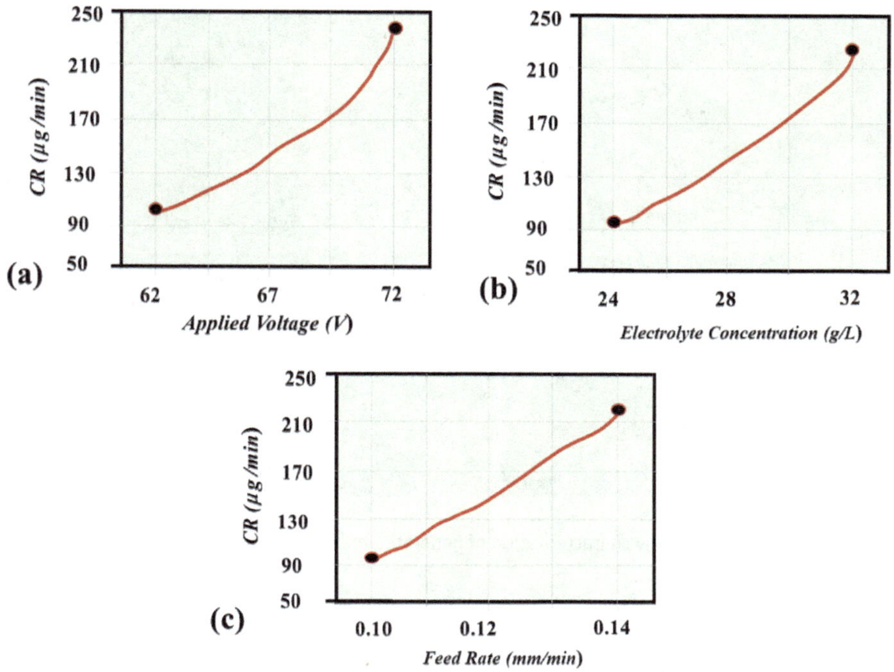

Fig. 2.3: Influence of process parameters on cutting rate (CR): (a) voltage versus CR, (b) electrolyte concentration versus CR, and (c) feed rate versus CR.

tool feed rates, the residence time of the tool over the workpiece surface becomes too short for stable gas film development, leading to irregular sparking and nonuniform energy transfer.

2.3.2 Influence of Machining Parameters on Depth of Cut

The electrolyte composition fundamentally governs the ECDM process dynamics during microhole fabrication, directly influencing both discharge characteristics and material removal mechanisms. Electrolyte in ECDM plays a dual role, i.e., it facilitates the development of H_2 bubbles through electrolysis and supports the creation of spark discharges at the tool-electrolyte interface. Electrolyte solution critically influences the performance by regulating both ionic transport phenomena and heat dissipation dynamics in the machining gap during machining. A low concentration reduces ionic conductivity, leading to delayed gas bubble formation and sparse sparking. Thus, the spark energy delivered to the workpiece surface was insufficient to cause deep material erosion, leading to shallow DOC. Figure 2.6(a–c) presents the effect of parameters on the DOC of the ECDM microhole. During the experimentation, the DOC of the microhole was varied from 735 to 1,390 μm, while increasing electrolyte concentration

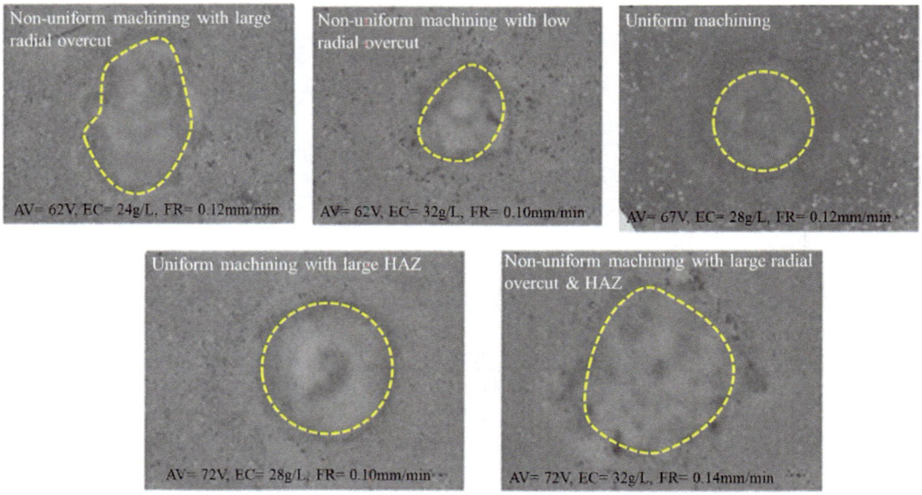

Fig. 2.4: Microscopic morphology characterization of generated microhole across different processing conditions.

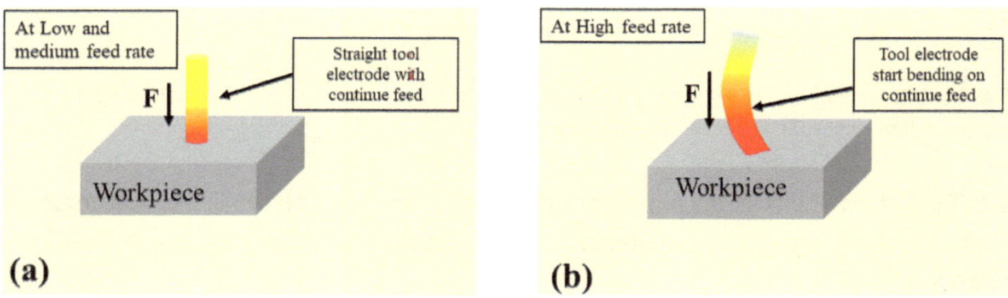

Fig. 2.5: Tool position: (a) at lower feed rate and (b) at higher feed rate.

from 24 to 32 g/L. The graph shown in Fig. 2.6(b) also indicates that the DOC varies nonlinearly with the electrolyte concentration. However, at higher concentration, this leads to excessive gas generation and unstable gas film behavior. The frequent bursting of bubbles interrupts spark formation and results in inefficient energy transfer. Additionally, high ionic strength can cause rapid passivation and surface fouling at the tool tip, reducing discharge efficiency.

Further, the applied voltage was the primary driver of electrolysis and spark generation during the ECDM process. It directly influences the threshold energy required for dielectric breakdown of the gaseous layer and subsequent spark energy. At lower applied voltages, the electric field strength was insufficient to initiate stable discharge across the gas film and the spark energy released per discharge was low, and therefore, the thermal erosion rate and CR were minimal, which leads to minimum DOC of microholes. As the applied potential rises, the dielectric breakdown of the gaseous

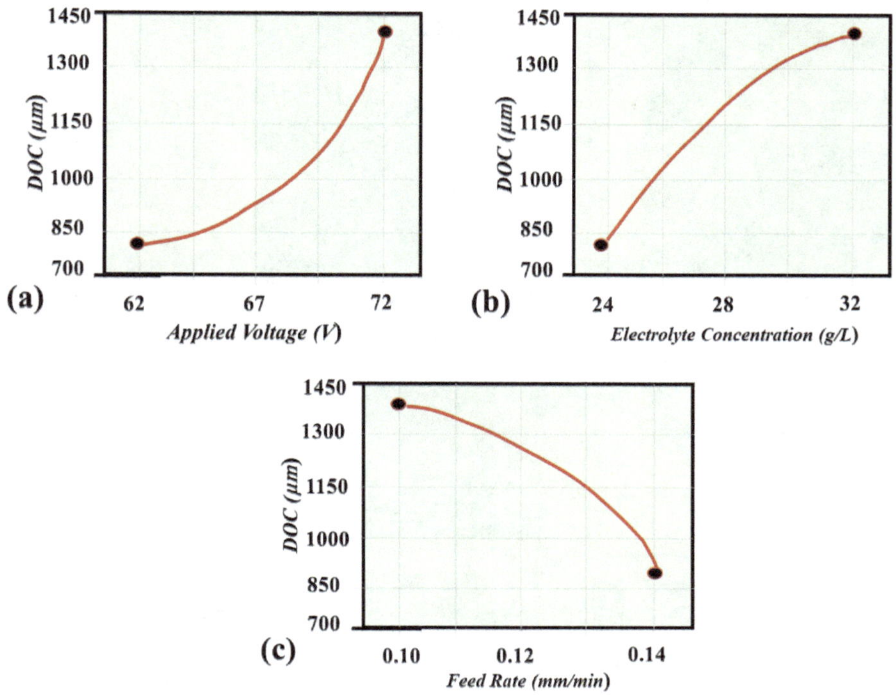

Fig. 2.6: Influence of process parameters on depth of cut (DOC): (a) voltage versus DOC, (b) electrolyte concentration versus DOC, and (c) feed rate versus DOC.

layer occurs regularly, producing consistent and intense sparks. These high-temperature sparks melt and vaporize the material, significantly enhancing DOC. At higher voltages, the spark energy behavior becomes uncontrolled, which leads to excessive heat generation in the machining region, and thermal cracking in the Macor ceramic due to its brittle nature and low thermal conductivity. The surface metrology of microholes machined on Macor dental ceramic under varying parameters was analyzed using SEM and EDX, as depicted in Fig. 2.7. SEM revealed differences in surface morphology, such as microcracks, recast layers, and debris, influenced by process conditions. EDX analysis identified elemental changes on the machined surface, indicating thermal-chemical interactions and material transfer. These analyses confirmed that optimized parameters yield superior surface integrity, essential for microfabrication in biomedical and electronic applications. Additionally, it can cause localized microfractures and lead to poor dimensional accuracy of microholes. While the DOC may still increase, the quality and integrity of the microhole may deteriorate. The behavior of applied voltage versus DOC of microholes is illustrated in Fig. 2.6(a).

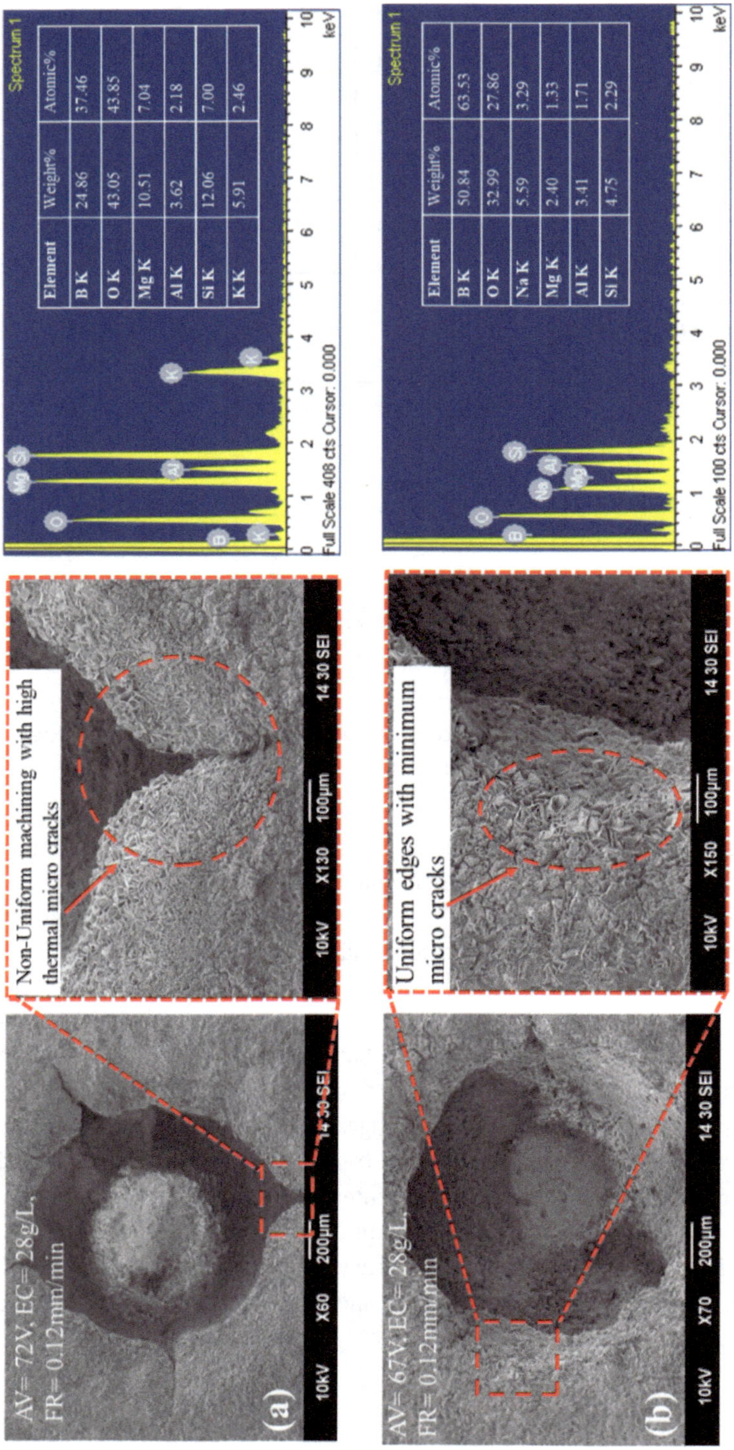

Fig. 2.7: SEM-EDX analysis of microholes machined under varying process parameters.

Moreover, the thermal energy imparted to the workpiece was also varying with the tool feed rate. A lower tool feed allows prolonged energy to accumulate at the machining region, promoting better melting and vaporization of material. This prolonged thermal exposure enhances material removal, resulting in higher DOC. However, prolonged dwell duration can potentially expand the HAZ while compromising dimensional precision. Experimental observations revealed an inverse relationship between feed rate and penetration depth – increasing the feed rate from 0.1 to 0.14 mm/min resulted in a corresponding reduction in cut depth from 1,410 to 890 μm. The variation of feed rate with DOC is shown in Fig. 2.6(c). The graph reveals that at higher feed rates, the reduced dwell time limits heat accumulation in the machining region, which adversely affects the depth of material penetration and overall cutting efficiency. As a result, fewer sparks interact with a given area of the workpiece, diminishing thermal damage. However, the process also becomes unstable at elevated feed speeds due to inadequate gas film formation, which is critical for maintaining consistent spark generation and machining performance.

2.4 Conclusions

The work reported in this chapter comprehensively evaluated the machinability of Macor dental ceramic through the ECDM process, emphasizing the influence and optimization of critical input parameters, such as applied voltage, electrolyte solution, and tool feed rate on micromachining performance. Experimental results demonstrated that increasing the various input variables within the selected range significantly impacted the CR, primarily by enhancing the electrochemical discharge frequency and energy density at the tool-workpiece interface. The maximum CR of 233.5 μg/min was achieved at an applied voltage of 72 V, electrolyte solution of 32 g/L, and a feed rate of 0.12 mm/min. Similarly, the greatest microhole depth of 1,410 μm was recorded at 72 V, 28 g/L, and 0.10 mm/min, respectively. These results affirm the synergistic role of controlled energy input, ionic conductivity, and tool dwell time in facilitating efficient thermal material removal. Morphological investigations using an optical microscope and SEM revealed the evolution of surface integrity with varying parametric settings. Increased levels of voltage and concentration were associated with pronounced thermal effects, including localized melting, formation of microcracks, and expansion of the thermally HAZ, particularly in regions of unstable spark generation. Conclusively, this research provides a valuable reference framework for working in micromachining of hard, brittle, and nonconductive materials, particularly in dental and biomedical applications. By establishing process-structure relationships and following optimal parameter windows, the outcomes can guide future process development, modeling, and hybridization efforts in ECDM and other nontraditional machining techniques for processing ceramics and other materials.

List of Acronyms

Acronyms	Description
ECDM	Electrochemical discharge machining
MRR	Material removal rate
CR	Cutting rate
DOC	Depth of cut
OC	Overcut
SEM	Scanning electron microscopy
EDS	Energy-ispersive X-ray spectroscopy
TW	Tool wear
GRA	Gray relational analysis
PMMA	Polymethyl methacrylate
MADM-R	Multiattribute decision-making using ranking
DC	Direct current
DOE	Design of experiment
NaOH	Sodium hydroxide

References

[1] Singh RP, Singhal S. Rotary ultrasonic machining of macor ceramic: an experimental investigation and microstructure analysis, Mater Manuf Process. 2017;32(9):927–939. doi:10.1080/10426914.2016.1198033.

[2] Singh RP, Singhal S. An experimental study on rotary ultrasonic machining of Macor ceramic, Proc Inst Mech Eng Part B J Eng Manuf. 2018;232(7):1221–1234. doi:10.1177/0954405416666897.

[3] Bindu Madhavi J, Hiremath SS. Machining and characterization of channels and textures on quartz glass using μ-ECDM process, Silicon. 2019;11(6):2919–2931. doi:10.1007/s12633-019-0083-6.

[4] Vaishya R. et al. Finite element modeling of quartz material for analyzing material removal rate in ECDM process, Int J Interact Des Manuf. 2023;17(5):2453–2459. doi: 10.1007/s12008-022-01037-2.

[5] Hui Y, Na ZL, Sun BM, Sun XD, Chen JL. A novel laser scanning method for metallization of alumina substrate by copper, Rare Met. 2023;42(3):844–852. doi:10.1007/s12598-022-02157-0.

[6] Luo Y. et al. Material removal mechanism of alumina ceramics with powder-mixed electrochemical discharge-drilling hybrid machining, Ceram Int. 2025; (In Press), doi:10.1016/j.ceramint.2025.01.152.

[7] Kumar S, Dvivedi A. On effect of tool rotation on performance of rotary tool micro- ultrasonic machining, Mater Manuf Process. 2019;34(5):475–486. doi:10.1080/10426914.2018.1512130.

[8] Singh M, Singh S. Electrochemical discharge machining: A review on preceding and perspective research, Proc Inst Mech Eng Part B J Eng Manuf. 2019;233(5):1425–1449. doi:10.1177/0954405418798865.

[9] Sambathkumar S, Arunagirinathan RS. Fabrication of micro-mechanical planar cantilever beam-mass structures on quartz substrates using an ECDM process, Proc Inst Mech Eng Part B J Eng Manuf. 2022;236(6–7):668–679. doi:10.1177/09544054211040617.

[10] Zhang S, Zhou J, Hu G, Wang L, Xu Y. Process characteristics of electrochemical discharge machining and hybrid methods: a review, Int J Adv Manuf Technol. 2023;129(5–6).

[11] Harugade M, Waigaonkar S, Dhawale N. A novel approach for removal of delaminated fibers of reinforced composites using electrochemical discharge machining, Proc Inst Mech Eng Part B J Eng Manuf. 2021;235(12):1949–1960. doi:10.1177/09544054211014483.

[12] Verma S, Satsangi PS, Chattopadhyay KD. Enhancing process capabilities of electric discharge machining for nonconductive ceramics, Proc Inst Mech Eng Part C J Mech Eng Sci. 2020;234 (12):2402–2416. doi:10.1177/0954406220905867.
[13] Wang T, Liu Y, Wang K, Lv Z. Investigation on a sustainable composite method of glass microstructures fabrication—Electrochemical discharge milling and grinding (ECDM-G), J Clean Prod. 2023;387:135788. doi:10.1016/j.jclepro.2022.135788.
[14] Arya RK, Appalanaidu B, Dvivedi A. Parametric investigation on an adaptive tool feeding system for ECDM, Int J Interact Des Manuf. 2023;17:–. doi:10.1007/s12008-023-01227-6.
[15] Tiwari AK, Panda SS. Optimization of process parameters in ECDM machining using Taguchi-based grey relation analysis, Measurement. 2023;216:112971. doi:10.1016/j.measurement.2023.112971.
[16] Sandhu IS, Maurya SK, Manna A. Experimental investigation on ECDM parameters during µ-drilling of fabricated Zn/(Ag + Fe)-MMC for biodegradable application, J Braz Soc Mech Sci Eng. 2023;45 (8):1–13. doi:10.1007/s40430-023-04296-6.
[17] Bellubbi S, N S, Mallick B. Multi-response optimization of ECDM process parameters for machining of microchannel in silica glass using Taguchi–GRA technique, Silicon. 2022;14(8):4249–4263. doi:10.1007/s12633-021-01167-4.
[18] Antil P. Modelling and multi-objective optimization during ECDM of silicon carbide reinforced epoxy composites, Silicon. 2020;12(2):275–288. doi:10.1007/s12633-019-00122-8.
[19] Bhargav KVJ, Balaji PS, Sahu RK, Katiyar JK. Multi-response optimization and effect of tool rotation on micromachining of PMMA using an in-house developed µ-ECDM system, CIRP J Manuf Sci Technol. 2022;38:473–490.
[20] Kumar R, Singh RP. On the machinability and formation of gas film thickness in ECDM of silica-based Pyrex glass, J Electrochem Soc. 2024;171(9):93510. doi:10.1149/1945-7111/ad7bf3.
[21] Appalanaidu B, Dvivedi A. On controlling of gas film shape in electrochemical discharge machining process for fabrication of elliptical holes, Mater Manuf Process. 2021;36(5):558–571. doi:10.1080/10426914.2020.1854464.
[22] Ma N, Chen X, Zhang A, Zhou J. Experimental study on immersed rotating cylindrical electrode in electrochemical discharge machining (ECDM), Mater Manuf Process. 2025;00(00):1–15. doi:10.1080/10426914.2025.2462845.
[23] Kumar R, Singh RP, Singh S. Enhancing the machining performance during micro-channel fabrication on quartz ceramic with rotary ECDM process, Mater Manuf Process. 2025;00(00):1–18. doi:10.1080/10426914.2025.2462832.
[24] Ben Mhahe F, Zhang Y, Chen C, Umoren WI. Investigation of the influence of different types of electrolytes on the performance of the electrochemical discharge machining process during micromachining of molybdenum, Int J Electrochem Sci. 2024;19(7): doi:10.1016/j.ijoes.2024.100647.
[25] Kumar M, Vaishya RO, Suri NM, Manna A. An experimental investigation of surface characterization for zirconia ceramic using electrochemical discharge machining process, Arab J Sci Eng. 2021;46 (3):2269–2281. doi:10.1007/s13369-020-05059-4.

Yash Yalamalli, Vibha Sastry H. K., Priyadarshan V., Anshul Agrawal,
Siddhartha Kar*

3 Hybrid Electrodischarge Machining Processes: Fundamentals and Developments

Abstract: Electrodischarge machining (EDM) is a prominent nontraditional thermal machining technique used for shaping electrically conductive and hard-to-machine materials. The process operates on the principle of thermal erosion caused by electrical discharges between the tool electrode and the workpiece, which eliminates chatter and vibration by avoiding direct mechanical interaction. Despite its advantages, traditional EDM faces limitations, including low machining efficiency, reduced accuracy, and suboptimal surface quality. To overcome these challenges, various enhancement techniques have been developed, including powder-mixed EDM, cryogenic-assisted EDM, and magnetic levitation EDM, which focus on equipment modification or the integration of auxiliary systems. Beyond these improvements, EDM has also been hybridized with other machining principles to form processes that offer synergistic benefits, including electrodischarge grinding, ultrasonic-assisted EDM, magnetic field-assisted EDM, and laser-assisted EDM. These hybrid processes combine the strengths of multiple methods to achieve higher efficiency, improved surface finishes, and the capability to machine complex geometries. This chapter examines the fundamental mechanisms and equipment configurations of these hybrid EDM processes, highlighting their advantages, limitations, and diverse industrial applications. Furthermore, it reviews current research trends with advancements in hybrid EDM and outlines future directions for innovation and development in this dynamic field.

Keywords: hybrid machining, EDM, vibration, magnetic field, laser, electrodischarge grinding

3.1 Introduction

Electrodischarge machining (EDM) is a nontraditional material removal process that operates through a series of rapid electrical discharges between a conductive workpiece and a tool electrode, all submerged in a dielectric fluid [1]. These discharges occur within a narrow voltage gap separating the tool electrode from the conductive workpiece. The intense heat produced by each discharge locally vaporizes small volumes of the workpiece material. The dielectric fluid, continuously flushed through the interelectrode gap (IEG), cools the area and removes the eroded debris from the machining region. The foundational principle behind EDM dates back to the 1770s,

 | https://doi.org/10.1515/9783111704456-003

when Joseph Priestley first observed the erosive effect of electrical discharges on metal surfaces [2]. Later, B. R. Lazarenko and N. I. Lazarenko pioneered a controlled EDM process using a resistance-capacitance power supply to machine hard metals by vaporizing material, a system widely adopted in the 1950s. However, the early implementations of EDM technology were limited by a lack of precision and control, making the process unsuitable for industrial applications at the time [3]. A significant breakthrough occurred in the mid-1970s with the commercial development of wire electrodischarge machining (WEDM), which enabled greater control, accuracy, and repeatability. This advancement established EDM as a practical and reliable manufacturing process. By the mid-1980s, EDM technology had been integrated into standard machine tool systems, leading to its widespread adoption [4]. Today, EDM plays a vibrant role in the manufacturing of tools, dies, and molds, particularly in processing hard-to-cut materials like hardened tool steels. EDM's capability to machine complex geometries with exceptional accuracy has made it an indispensable technique in modern manufacturing environments [5].

3.1.1 Working Principle

EDM is an advanced machining technique that erodes material by transforming electrical energy into highly concentrated thermal energy, a principle known as thermoelectric energy conversion [1]. Figure 3.1 illustrates the working principle of EDM, where the tool acts as the negative electrode and the workpiece as the positive electrode, both submerged in a dielectric fluid with a precisely controlled IEG, typically ranging from 0.01 to 0.5 mm [6]. Pulsed voltages, typically ranging from 20 to 120 V at a frequency of around 5 kHz, are applied across the electrodes [7]. As the IEG narrows, the electric field intensity builds up to the point where the dielectric fluid breaks down, initiating a spark discharge. This discharge forms a highly ionized plasma channel, generating localized temperatures of several thousand degrees Celsius. The extreme heat melts and vaporizes a minute volume of the workpiece material, leading to the formation of microcraters on its surface [7, 8]. The molten material is then rapidly quenched and solidified by the surrounding dielectric fluid, forming fine, spherical debris particles, which are carried away by the fluid's circulation [9]. The dielectric fluid, commonly kerosene or deionized (DI) water, plays a decisive role throughout the machining process. Initially, it functions as a nonconductive medium, preventing premature discharges until the breakdown voltage is reached [9]. Once a spark occurs, the fluid functions as a coolant, dissipating heat from both the tool and the workpiece, while simultaneously flushing away eroded material to maintain a clean and stable IEG [8, 10]. After each discharge, the fluid cools and deionizes, regaining its insulating properties in preparation for the next pulse. The effectiveness of dielectric flushing is vital for maintaining process efficiency, avoiding short circuits, and achieving a consistent material removal rate (MRR). Several process parameters significantly impact the quality and performance of EDM. These include pulse duration (T_{on}), discharge energy (DE), voltage level, and the physical properties of the dielectric fluid, all of which directly affect surface finish, MRR, and machining stability [8]. One of the primary advantages of EDM is its ability to machine any electrically conductive material regardless of hardness, owing to its entirely noncontact mechanism. This eradicates mechanical stresses and deformation, making EDM particularly suitable for intricate, delicate, or high-precision components.

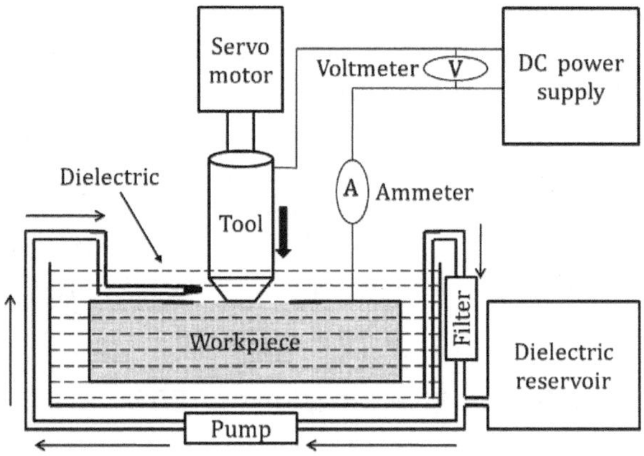

Fig. 3.1: Setup of electrodischarge machining (EDM).

3.1.2 EDM Process Parameters

EDM performance and efficiency are governed by a complex interaction of electrical, mechanical, thermal, and material-related parameters. These factors influence critical outputs like MRR, surface roughness (SR), dimensional accuracy, and tool wear (TW). A deep understanding of these parameters is critical for effective process control and optimization.

One of the most influential parameters is DE, which determines the intensity of each spark and directly affects both MRR and SR. Higher DE typically leads to increased MRR, but may degrade surface integrity and accelerate TW [11–13]. Voltage and discharge current (I_p) also play vital roles in controlling spark formation and energy input. Higher voltage facilitates dielectric breakdown over larger IEG, while increased I_p intensifies the spark, enhancing erosion but potentially compromising accuracy [14]. The T_{on} and pulse interval (T_{off}) are critical for controlling the duration and frequency of discharges. Longer T_{on} increases the energy delivered per pulse, thereby improving MRR, while adequate T_{off} allows dielectric recovery and debris removal, promoting spark stability [15]. The duty cycle, expressed as the proportion of T_{on} to the complete discharge cycle duration ($T_{on} + T_{off}$), influences the average power delivered and overall thermal load on the system. The IEG, which refers to the gap maintained separating the tool electrode from the workpiece, is another key factor affecting electric field strength and spark initiation. This IEG is precisely regulated by a servo mechanism to ensure stable discharges and consistent material removal [16–18]. Dielectric fluid properties significantly impact process stability and efficiency. The fluid acts as an insulator before breakdown, then assists in cooling and flushing debris during and after the spark. Its viscosity, dielectric strength, and flushing effectiveness directly affect crater formation, spark stability, and avoidance of secondary discharges [19, 20]. Flushing efficiency is particularly important in ensuring that eroded particles are promptly removed from the IEG. Inadequate flushing may result in arcing, short-circuiting, and deteriorating the machined surface [21, 22]. Material characteristics also influence EDM behavior. TW is a critical consideration, as it affects machining precision and tool life [23]. The choice of electrode material, based on its melting point, thermal conductivity, and erosion resistance, determines its suitability for specific applications [24]. Additionally, workpiece material properties, including melting point, thermal conductivity, and electrical resistivity, govern how the material responds to DE, thereby affecting the ease and efficiency of machining [25].

3.1.3 Advantages, Disadvantages, and Applications of EDM

EDM offers unique advantages and specialized capabilities but also presents certain limitations. Understanding its strengths, drawbacks, and application areas is essential for selecting it as an appropriate manufacturing process. EDM can effectively process all electrically conductive materials regardless of their hardness, making it particularly effective for processing advanced engineering materials like hardened steels, titanium, and carbides [26]. It allows the creation of complex geometries and fine details that are challenging with conventional methods. Being a noncontact process, EDM avoids the introduction of mechanical stresses and eliminates issues related to chatter and vibrations during machining [2]. It is especially effective for drilling small, deep holes with high precision [27]. EDM is limited to conductive materials and cannot be used on nonconductors like ceramics or plastics without specialized techniques [28]. TW during machining affects dimensional accuracy and requires compensation [23]. Elevated temperatures may lead to the formation of a heat-affected zone (HAZ), which modifies the surface characteristics [29]. The use of dielectric fluids adds to the cost and environmental concerns, and the process can involve high power consumption and machine investment [20]. EDM is extensively employed in tool and die manufacturing to produce molds and dies for processes like injection molding, stamping, forging, and die casting [30]. In aerospace, it is applied for machining turbine blades, high-temperature alloys, and structural components with tight tolerances [31]. The automotive industry uses EDM for engine parts, dies, and prototype tooling [30]. In the medical field, it is used for surgical instruments, implants, and precision prosthetics [32]. It is also critical in micromachining for producing microfeatures in electronics, sensors, and MEMS devices [33].

3.1.4 Need for Hybridization of EDM

EDM has revolutionized the machining of hard and complex materials, surpassing many limitations of conventional techniques. However, its dependence on electrical conductivity restricts its use with insulating materials like ceramics, plastics, composites, and glass [9, 28]. To address this limitation and further enhance process capabilities, hybrid EDM technologies have been developed by integrating EDM with complementary machining or assistive techniques [34, 35]. These hybrid methods not only allow for processing nonconductive materials but also significantly improve key performance metrics. TW, though typically lower in standard EDM, can still compromise accuracy and tool life; hybrid processes have been shown to mitigate this issue effectively. Surface finish is also enhanced, as hybrid EDM can reduce SR and improve dimensional precision, often eliminating the need for postprocessing. Moreover, hybrid systems offer improved machining accuracy, especially for intricate and complex geometries that challenge conventional EDM approaches. In terms of productivity, hybrid EDM methods achieve higher MRR, meeting industrial demands for faster and more efficient processing [35]. Enhanced control over the material removal process also results in reduced waste and improved machining precision, making hybrid EDM a highly effective solution for modern manufacturing challenges.

3.1.5 Types of Hybrid EDM Processes

EDM-based hybrid machining processes are generally grouped into two main categories: primary and secondary hybrids, each developed to address the inherent limitations of conventional EDM such as low MRR, poor surface finish, and limited material compatibility [35]. Primary hybrids, also termed EDM-based primary hybrid machining processes, combine EDM with a fundamentally different machining technique to form a novel process. Notable examples include electrodischarge abrasive grinding (EDAG), where mechanical abrasion of conventional grinding is combined with spark erosion [36], and electrochemical discharge machining, which allows processing of electrically nonconductive materials by merging electrochemical and discharge principles [37]. Secondary hybrid processes, on the other hand, enhance the standard EDM mechanism through auxiliary energies or fields. These include vibration-assisted EDM (VAEDM), where ultrasonic vibration improves debris evacuation and spark stability [38]; laser-assisted EDM (LAEDM), which softens the workpiece to reduce TW and energy requirements [39]; magnetic field-assisted EDM (MFAEDM), which uses Lorentz forces to aid in debris flushing [40]; and the sequential EDM-electrochemical machining hybrid machining process, which combines EDM and electrochemical machining in a controlled sequence to optimize material removal and surface integrity [41]. These hybrid processes provide substantial improvements in machining efficiency, surface quality, and process stability, making them highly suitable for advanced manufacturing applications involving hard and complex materials. The scope of this chapter covers two major categories: (i) EDM combined with conventional machining, and (ii) EDM assisted by laser, vibration, and magnetic field.

3.2 EDM Combined with Conventional Machining

3.2.1 Electrodischarge Grinding (EDG)

Electrodischarge grinding (EDG) is a hybrid process that integrates electrical discharge and mechanical grinding principles to precisely machine materials that are both hard and difficult to grind, provided they are electrically conductive. Unlike traditional EDM, which uses a linearly moving electrode, EDG employs a rotating conductive grinding wheel, typically made of copper, brass, or mild steel, as the tool electrode [42, 43]. Material removal in EDG primarily takes place through spark erosion between the grinding wheel and the workpiece, both submerged in a dielectric medium like kerosene, transformer oil, or DI water. A typical configuration of the EDG setup, including the power supply, rotating conductive grinding wheel, dielectric system, and workpiece arrangement, is illustrated in Fig. 3.2. When a breakdown voltage ionizes the dielectric, a plasma channel forms and enables spark generation, producing localized temperatures between 8,000 and 20,000 °C, resulting in micromelting and vaporization of the electrodes. A servo-controlled IEG (typically 0.013–0.075 mm) and pulsed direct current (DC) power supply (30–400 V, 30–100 A, 2–500 kHz) regulate spark intensity and consistency [3]. The servo mechanism also compensates for wheel wear by dynamically adjusting the position of the wheel or workpiece. The wheel's rotation enhances dielectric flushing, efficiently removing molten debris and maintaining spark stability; the fluid is often filtered and recirculated to ensure a clean machining zone.

In EDG, the performance outcomes, primarily MRR, tool wear rate (TWR), and surface quality, are strongly governed by a set of process parameters unique to the hybrid nature of the technique [44]. Key electrical parameters include T_{on}, T_{off}, and the duty cycle, which collectively influence spark energy and frequency. The rotational speed of the conductive grinding wheel is critical, as it enhances dielectric flushing and improves debris evacuation. Other influential factors include electrode polarity, which determines erosion concentration on the tool or workpiece, and the type and composition of dielectric fluid, including additives that improve ionization, cooling, and flushing. These EDG-specific parameters must be precisely controlled to influence the synergy of spark erosion and grinding action for machining hard, con-

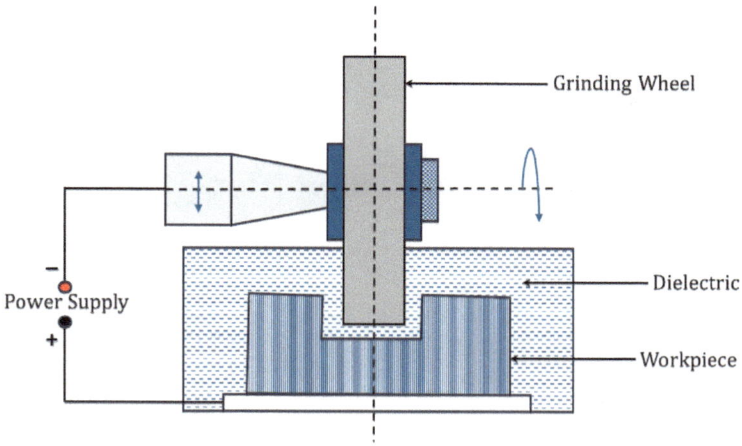

Fig. 3.2: Setup of electrodischarge grinding (EDG).

ductive materials with high accuracy. Compared with EDM, EDG is more productive, especially when machining extremely hard materials, where improvements of two to three times over conventional grinding are possible [45].

3.2.2 Electrodischarge Abrasive Grinding (EDAG)

EDG can be performed with or without abrasive grit embedded in the grinding wheel. When abrasives are present, the process is referred to as EDAG [46], which enhances both material removal and surface finish. The abrasives embedded in the wheel contribute to mechanical abrasion in addition to the material removal caused by electrical discharges between the electrodes [47]. A typical configuration of the EDAG setup, illustrating both the abrasive action of the grits and the electric sparks occurring at the wheel-workpiece interface, is illustrated in Fig. 3.3. EDAG has evolved into variants such as cut-off, face, and surface grinding, broadening its applicability for hard, thermally resistant, and geometrically complex materials [48]. The IEG, typically maintained within microns by a servo system, is vital for stable spark generation and consistent erosion. The protrusion height of abrasive particles from the wheel surface also affects the cutting action and surface finish, especially in EDAG [49]. For effective machining, protrusion height should be higher than IEG [50]. Additionally, normal and tangential forces generated during the wheel-workpiece interaction affect grain dislodgement and influence the combined thermal-mechanical removal mechanisms [51, 52].

The EDAG process offers several advantages over conventional machining techniques, particularly when dealing with hard-to-machine materials. It enables improved surface finish with finesse comparable to abrasive finishing methods, yet operates at significantly faster rates. The integration of EDM assistance in grinding not only en-

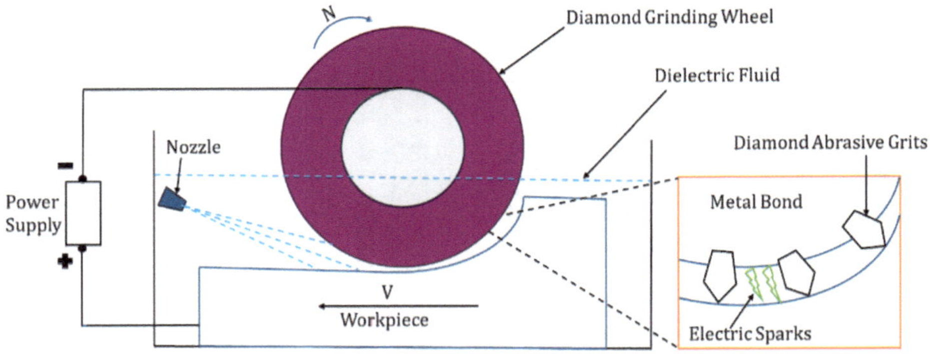

Fig. 3.3: Setup of electrodischarge abrasive grinding (EDAG).

hances the performance of grinding wheels but also extends their application range by softening and vaporizing the work material [53]. A key benefit of EDAG is its self-dressing capability, wherein electrical discharges concurrently erode both the workpiece and the grinding wheel, allowing continuous dressing without interrupting the machining process, thereby enhancing productivity [54, 55]. However, EDAG also presents notable drawbacks. These include the risk of unwanted erosion and overcutting, poor surface quality at high MRR, and elevated power consumption due to the concurrent use of EDM and grinding operations [56]. Furthermore, the surface finish produced by EDAG is often inferior to that of conventional grinding, necessitating additional finishing steps [54].

A comparative analysis of EDM and EDAG is presented in Tab. 3.1, which highlights the superior performance of EDAG over traditional EDM in key performance

Tab. 3.1: Comparison between EDM and EDAG.

Criteria	**Electrodischarge machining (EDM)**	**Electrodischarge abrasive grinding (EDAG)**
Surface and subsurface characteristics	Contains recast layer with cracks and voids [57]	Recast layer removed or uniform; minimal defects [57]
Surface roughness (Ra)	2–3.5 µm [57]	As low as 0.04 µm [57]
Material removal rate	Low, especially for ceramics [42]	Higher due to hybrid action [42]
Tool/electrode wear	High relative electrode wear ratio [42]	Lower relative electrode wear ratio [42]
Surface integrity	Microcracks, thermal spalling, altered composition [57]	Uniform microstructure, fewer cracks [57]
Environmental impact	Generates harmful gases; needs filtration [42]	Environmentally friendly with water-based fluids [42]

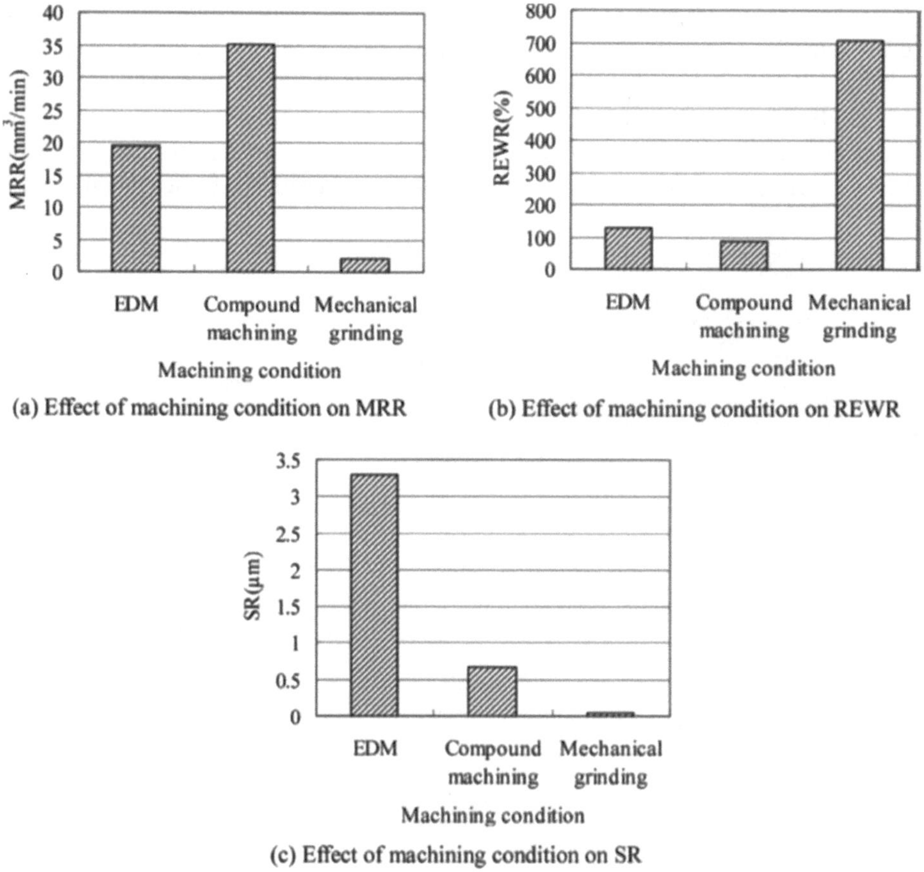

Fig. 3.4: Comparison of experimental results after machining SiC ceramic with EDM, compound machining (EDAG), and mechanical grinding [42] (with kind permission from Taylor & Francis).

metrics. Figure 3.4 quantifies the MRR, relative electrode wear ratio (REWR), and SR for EDM, EDAG, and traditional grinding [42]. As shown in Fig. 3.4(a), the MRR is significantly higher in compound machining, indicating improved machining efficiency. Figure 3.4(b) demonstrates that the REWR is substantially reduced in EDAG compared to EDM, reflecting enhanced tool life and cost-effectiveness. Furthermore, Fig. 3.4(c) shows a marked improvement in SR, where EDAG achieves a much smoother surface than EDM and closely approaches the finish quality of mechanical grinding. These observations align with the tabulated comparison, reinforcing the advantages of EDAG in achieving higher productivity, better surface quality, and lower TW, making it a promising hybrid approach for machining hard materials like ceramics.

Figure 3.5 presents micrographs of Al_2O_3-SiC_W-TiC surfaces machined by EDM and EDAG [57, 58]. After EDM at a low I_p of 3 A, the surface appears rough and uneven (Fig. 3.5(a)). At a higher I_p of 5 A, significant thermal damage is observed, including globules and micropores (Fig. 3.5(b)), primarily due to intense energy discharges and recast layer deposition. In contrast, the surface produced by EDG at the same I_p of 5 A appears noticeably smoother and more uniform (Fig. 3.5(c)). High-magnification images confirm minimal voids and the absence of cracks (Fig. 3.5(d)), highlighting the synergistic effect of electrical discharge and mechanical abrasion in EDG, which removes the recast layer before it solidifies, resulting in improved surface integrity.

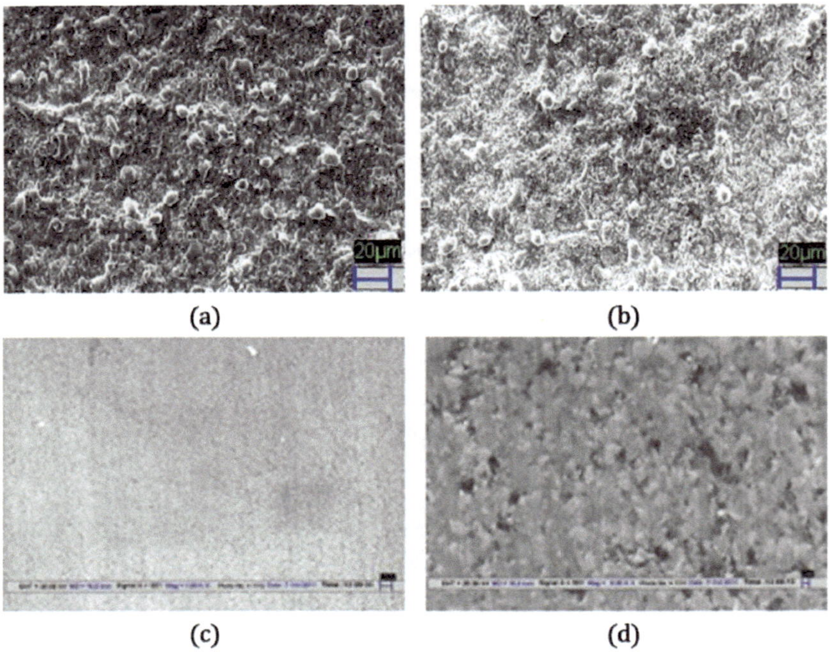

Fig. 3.5: SEM images of surface characteristics of Al_2O_3-SiC_w-TiC machined via EDM at a duty cycle of 0.72, gap voltage of 70 V, and T_{on} of 200 µs: (a) I_p = 3 A; (b) I_p = 5 A [58] (with kind permission from Elsevier); micrograph of Al_2O_3-SiC_w-TiC after EDAG at I_p = 5 A, duty cycle = 0.56, T_{on} = 200 µs: (c) showing surface texture; (d) showing voids at 6k magnification [59] (with kind permission from Taylor & Francis).

3.2.3 Electrodischarge Diamond Grinding (EDDG)

Electrodischarge diamond grinding (EDDG) is an advanced hybrid machining technique that integrates the principles of EDM with diamond-impregnated tools [60]. It is especially well suited for processing hard and brittle materials like ceramics, composites, and superalloys, which are grueling to machine conventionally. EDDG utilizes the high thermal conductivity and hardness of diamond to enable precise material removal while minimizing TW and thermal damage [61]. In this process, a diamond-impregnated grinding wheel acts as the electrode, generating electrical discharges in a dielectric fluid that melt and vaporize material. The diamond particles embedded in the grinding wheel do not directly conduct electricity but facilitate the generation of electrical discharges by maintaining a small IEG between the wheel and the workpiece. In addition to material erosion owing to sparking during T_{on}, the abrasive action of diamond grits also remove material during T_{off} [60]. Their exceptional hardness and thermal conductivity of diamond help dissipate heat effectively, preserve abrasive grain integrity, and significantly reduce TW, thereby enhancing tool life and minimizing downtime. Further, EDDG has also been highly effective in achieving superior surface integrity by significantly minimizing thermal damage and recast layer formation. Figure 3.6 presents scanning electron microscopy (SEM) based micrographs of machined surfaces under different input conditions [60]. At high DE, a thick recast layer is observed (Fig. 3.6(a)), indicating extensive thermal loading. When operated at moderate energy (Fig. 3.6(b)), the recast layer becomes noticeably thinner, reflecting improved control over heat input. Under low DE conditions (Fig. 3.6(c)), the recast layer is nearly eliminated (2 μm), demonstrating the process's ability to produce surfaces with minimal subsurface damage.

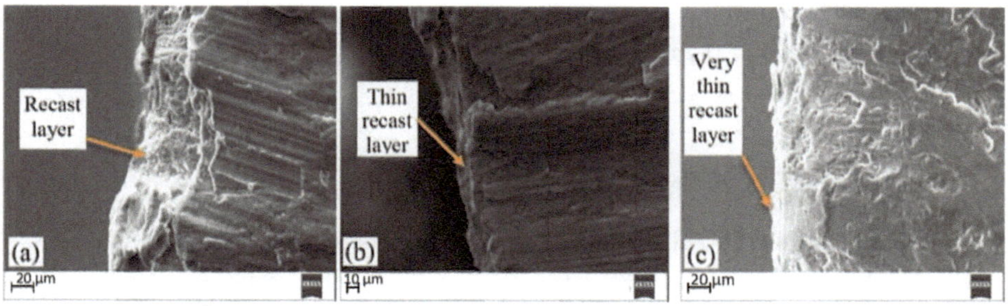

Fig. 3.6: SEM images depicting the recast layers on surfaces machined by EDDG at different discharge energies: (a) high, (b) medium, (c) low [60] (with kind permission from Taylor & Francis).

3.2.4 Latest Developments in EDG, EDAG, and EDDG

Future directions in EDG, EDAG, and EDDG aim to advance material removal efficiency, enhance surface integrity, and promote process sustainability. Innovations like slotted or hollow rotating disk electrodes improve debris flushing, spark stability, and uniformity of material removal, addressing key limitations of conventional setups [62]. Additionally, emphasis on high thermal conductivity and wear-resistant tool materials aims to extend tool life and ensure consistent performance. Wire EDG, a variant of EDG employing a continuously fed wire electrode, is gaining prominence for its capability to fabricate microfeatures with exceptional dimensional accuracy and surface integrity [63]. It has demonstrated potential in applications requiring tight tolerances, with ongoing research exploring its adaptability to diverse materials and complex geometries [64, 65].

Future developments in EDDG are expected to enhance bond matrix conductivity and grit retention, thereby improving the wheel's mechanical and thermal performance [66]. Such advancements are critical for achieving consistent surface finish and reduced TW, especially in high-precision or high-load applications. Beyond tool design, the incorporation of powder-mixed dielectrics is emerging as a key enhancer of EDG, EDAG, and EDDG performance. Conductive powders such as graphite or silicon improve spark distribution, enlarge the IEG, and contribute to better surface finish and reduced TW [67, 68]. These environments stabilize the discharge channel and promote uniform material removal, with future research likely to focus on optimizing powder type, concentration, and particle size for improved process control. Hybrid machining approaches, such as integrating EDG, EDAG, and EDDG with ultrasonic assistance, magnetic fields, or cryogenic cooling, are gaining attention for mitigating thermal effects and enhancing surface integrity. These techniques improve debris removal and reduce HAZ. Concurrently, the use of smart control systems, real-time monitoring, and artificial intelligence-driven discharge optimization is poised to advance automation and process reliability [69]. Collectively, these innovations can advance grinding precision and lay the groundwork for sustainable, intelligent EDG systems.

3.3 EDM Assisted by Laser, Magnetic Field, and Vibrations

3.3.1 Laser-Assisted Electrodischarge Machining (LAEDM)

Laser-assisted machining combines laser heating with conventional or nontraditional processes to improve machinability, particularly for hard-to-machine materials. A laser emits a coherent, high-intensity beam ideal for material processing. Common lasers include Nd:YAG and CO_2, with the former offering better material absorption and the latter delivering higher power for fast cutting [70]. In machining, lasers are typically used to remove material through melting or vaporization. However, in LAEDM, the laser preheats the workpiece to reduce electrical resistance, enhance discharge efficiency, and improve MRR. This hybrid method offers improved surface quality, reduced TW, and greater efficiency compared to traditional EDM. Figure 3.7 displays the schematic setup of LAEDM, integrating an Nd:YAG laser and EDM system to enhance material removal efficiency and machining precision [71].

Various configurations have been developed to integrate Nd:YAG lasers with EDM for advanced micromachining. In one setup, EDM performs part fabrication while the laser enables precise microjoining for assembly [39, 72]. A sequential process com-

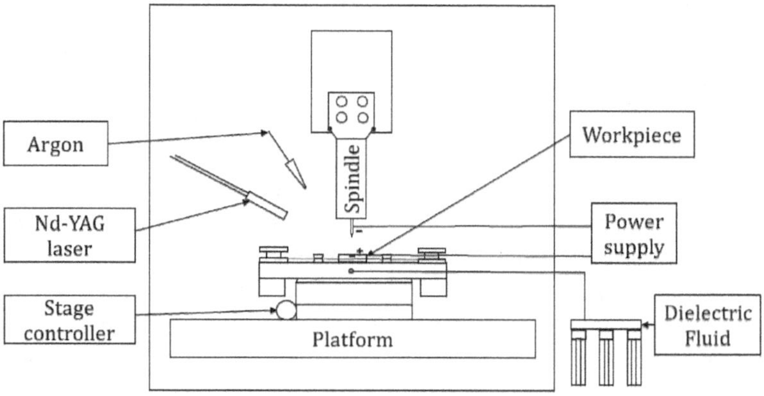

Fig. 3.7: Setup of laser-assisted EDM (LAEDM) [71] (with kind permission from Elsevier).

bines laser drilling of pilot holes, end milling for diffuser shaping, and microdrilling for final finishing, ideal for producing high aspect ratio inclined holes with minimal burr and recast layer [73]. Another configuration integrates ultrashort pulsed laser ablation with micro electrodischarge machining (μEDM) milling in a single machine, eliminating reclamping and combining the strengths of both techniques for efficient, high-precision microfeature fabrication [74].

LAEDM enhances the conventional EDM process by integrating laser energy to preheat the workpiece, guide discharge paths, and improve machining efficiency and precision. Initially, a focused laser beam raises the workpiece surface temperature, softening the material and reducing its effective melting point, thereby facilitating easier material removal during subsequent electrical-discharge events [75]. The laser can also ionize the dielectric fluid or surrounding air, creating a low-resistance plasma channel that aids in initiating and stabilizing the electrical discharge. Advanced laser beam profiles, such as Bessel or Airy beams, have been shown to guide the discharge path precisely, even around geometrical constraints or surface irregularities [76]. Once a high-voltage pulse is employed between the tool and the preheated workpiece, electrical breakdown occurs more easily, leading to controlled and localized material removal through melting and vaporization. The dielectric fluid serves a dual purpose by cooling the machining area and removing molten debris simultaneously [77]. Additionally, the laser's influence enhances discharge stability by mitigating disturbances caused by dielectric turbulence or debris accumulation, improving overall process reliability. In some configurations, the laser also assists in postmachining surface cleaning, helping reduce oxidation and contamination on the machined surface [78]. This hybrid technique synergizes the advantages of laser heating and electrical discharge, enabling superior machining performance, especially for difficult-to-machine materials.

EDM and laser machining each exhibit inherent limitations when used independently. Conventional and μEDM processes are often constrained by low MRR and significant TW, particularly at the electrode edges. In contrast, laser machining offers higher machining speeds, but it can introduce drawbacks such as HAZ and recast layers, especially when using longer T_{on} [70]. While EDM is known for delivering superior surface finish, it lacks the efficiency of laser-based processes. By integrating laser assistance with EDM, the complementary advantages of both methods can be harnessed, improving machining speed, reducing TW, and maintaining excellent surface integrity, thereby overcoming the individual limitations of each technique. A comparative summary of conventional EDM and LAEDM is presented in Tab. 3.2, highlighting the key differences in performance, limitations, and machining outcomes. Rashid et al. [79] compared conventional μEDM with the hybrid LAEDM process, demonstrating that LAEDM reduced machining time (MT) by 72.36%, TW by 90.38%, and short circuit occurrences by 97.78% (Fig. 3.8). These enhancements are attributed to the laser's ability to premachine the material, thereby reducing the EDM load and improving process stability without compromising hole quality. Figure 3.9 shows that the hybrid laser beam micromachining (LBMM)-μEDM process yields microholes with markedly better morphology than those produced by either pure LBMM or pure μEDM. The sequential process leverages the high-speed material removal of LBMM and the precision of μEDM, resulting in cleaner hole walls, minimal recast layer, improved circularity, and significantly lower TW and MT.

Tab. 3.2: Comparison between EDM and LAEDM.

Criteria	Electrodischarge machining (EDM)	Laser-assisted electrodischarge machining (LAEDM)
Material removal rate (mm^3/min) [80]	Slower: 0.001422	Faster: 0.002082
Taper angle [80]	High due to tool wear	Lower due to debris flushing
Surface finish [78]	Good but slow	Reduced HAZ enhances surface quality
Tool wear [78, 79]	High due to sparks	Up to 9× lower as laser softens material
Machining time [81]	Long	Reduced as bulk material is removed by lasers first.
Surface defects [82]	Minimal	Negligible
Stability [79]	Frequent short circuits	Minimal short circuits

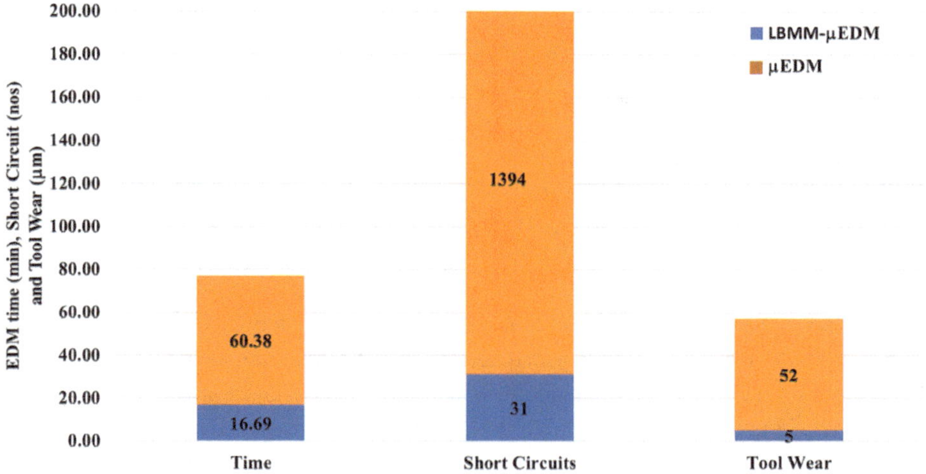

Fig. 3.8: Comparison of machining time, number of short circuits, and tool wear in LBMM-µEDM versus conventional µEDM [79] (with kind permission from Springer Nature).

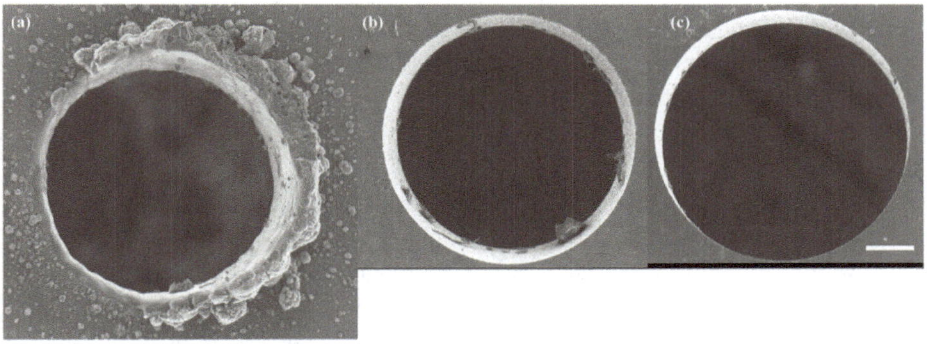

Fig. 3.9: Morphology of microholes produced by (a) LBMM, (b) µEDM, and (c) sequential LBMM-µEDM. Laser processing performed at 15.5 W power, 50 mm/s scan speed, and 20 kHz pulse frequency. Scale bar = 100 µm [79] (with kind permission from Springer Nature).

3.3.1.1 Latest Developments in Laser-Assisted Electrodischarge Machining

Recent advancements in LAEDM have led to hybrid techniques for improved precision and efficiency in machining complex microstructures. Shiu et al. [83] developed a cost-effective method combining laser micromachining with µEDM for the production of high-aspect-ratio micromolds used in lab-on-a-chip devices. Laser-cut 2D profiles on a brass sheet served as EDM masks, enabling high accuracy mold production as a scalable alternative to traditional lithography. Thermal processes like EDM, laser machining, and LAEDM often create HAZ, leading to microcracks and reduced fatigue life. To overcome this, Wälder and Richard [84] used picosecond laser machining with cold ablation, eliminating the HAZ and improving fatigue life by 40%. The method showed material sensitivity and holds promise for mold and die manufacturing requiring high surface integrity and durability. To address surface quality issues in selective laser melting (SLM), it has been integrated with µEDM for postprocessing. This hybrid approach significantly improves surface finish and accuracy, especially for Ti-6Al-4V parts. SR was reduced from 14.7 to 0.8 µm, and porosity to 0.17%, making it ideal for microimplants and precision components [85]. Furthermore, Chung et al. [86] developed a hybrid LAEDM process for machining carbon fiber reinforced plastic. A Ytterbium pulsed fiber laser removes the epoxy layer and defines geometry, followed by EDM finishing. This two-step method reduces taper, surface defects, and thermal damage while maintaining precision, proving effective for machining composite materials. Song et al. [87] developed a hybrid process combining SLM with WEDM for the fabrication of microgrippers featuring high-precision flexure hinges. This method reduced MT by 61% and increased fatigue life by over 237%, demonstrating the potential of LAEDM in enhancing precision, efficiency, and durability in micromanufacturing. Ma et al. [88] developed a hybrid technique combining picosecond laser, µEDM, and grinding to fabricate high-precision polycrystalline diamond microdrills. The laser enabled rapid shaping, EDM formed flutes, and grinding refined edges and removed defects. This LAEDM approach achieved superior tool quality and drilling performance, marking a significant advancement in microtool fabrication for hard materials.

3.3.2 Vibration-Assisted Electrodischarge Machining (VAEDM)

The low efficiency of material removal in EDM remains a significant concern [35]. One effective approach to overcome this is the integration of vibrations into the process, leading to VAEDM. Vibrations within the IEG induce cavitation effects in the dielectric fluid, and the collapse of these bubbles produces microjets that enhance evacuation of debris and molten material [89]. This improved flushing action increases dielectric circulation and reduces debris accumulation, thereby enhancing discharge stability and increasing the effective discharge ratio, both of which contribute to a substantial improvement in MRR [90, 91]. However, the inherently narrow IEG required for spark generation can lead to machining instability, resulting in surface defects, increased MT, and high TW. Inefficient debris evacuation during the micromachining of complex features often causes debris bridging between the electrodes (tool and workpiece), leading to short circuits and arcing, which degrade surface quality and machining efficiency. The improved flushing in VAEDM addresses these issues by minimizing debris concentration, suppressing arcing, and improving overall machining stability [92, 93]. This phenomenon is particularly advantageous when machining microslots and intricate contours. Figure 3.10 illustrates the schematic of a VAEDM setup, where vibrations, produced by a generator and transducer, can be employed to either of the tool or workpiece. Enhanced removal of molten material and debris also minimizes the formation of microcracks and microstructural damage on the machined substrate [94]. Studies have shown that VAEDM can be eight times more efficient than traditional EDM, with notable improvements in machining efficiency and dimensional accuracy of microfeatures [95].

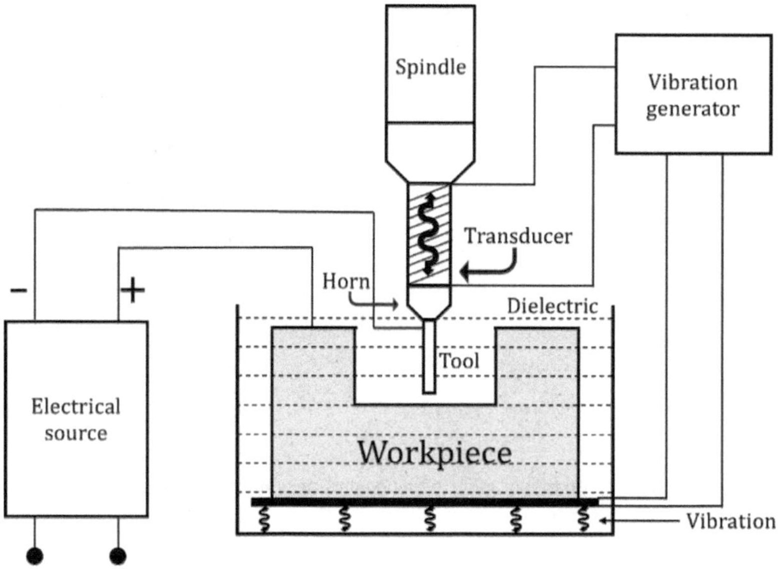

Fig. 3.10: Setup of vibration-assisted EDM (VAEDM).

VAEDM builds upon the conventional EDM process by introducing mechanical vibrations, typically imparted to either the tool or the workpiece, as displayed in Fig. 3.10 [35, 89, 92, 96, 97]. These vibrations, often in the ultrasonic or low-frequency range, help to shake the tool or workpiece. In traditional EDM, one of the major challenges is ensuring that the microscopic debris generated by each spark does not accumulate in the IEG, as this can destabilize the discharge and reduce efficiency. With vibrations, the electrode subtly moves in a controlled rhythm, constantly varying the IEG between itself and the workpiece. This motion not only enhances the flushing action of the dielectric fluid, allowing for smoother debris removal but also refreshes the spark zone, making discharges more consistent and reliable [98]. As a result, VAEDM tends to show noticeable improvements in MRR, surface quality, and tool life [91, 99]. Figures 3.11 and 3.12 compare the machined surfaces after EDM without and with the application of vibration to the workpiece, respectively [96]. Figure 3.11(a,b) reveals a rougher texture with more debris and spherical particles adhering to the surface, indicating inefficient chip removal. In contrast, Fig. 3.12(a,b) shows a cleaner surface with fewer adhered debris particles and reduced surface irregularities, attributed to improved flushing and more stable discharge due to vibration. The workpiece vibration enhanced dielectric circulation and debris ejection, resulting in superior surface quality, as evident in Fig. 3.12, compared to conventional EDM. It is a simple yet effective method of hybridizing EDM, especially valuable when working with hard-to-machine materials or precision features. A comparison of the essential criteria is presented in Tab. 3.3, indicating that VAEDM offers higher machining efficiency, reduced TW, improved surface quality, and greater cost-effectiveness compared to conventional EDM.

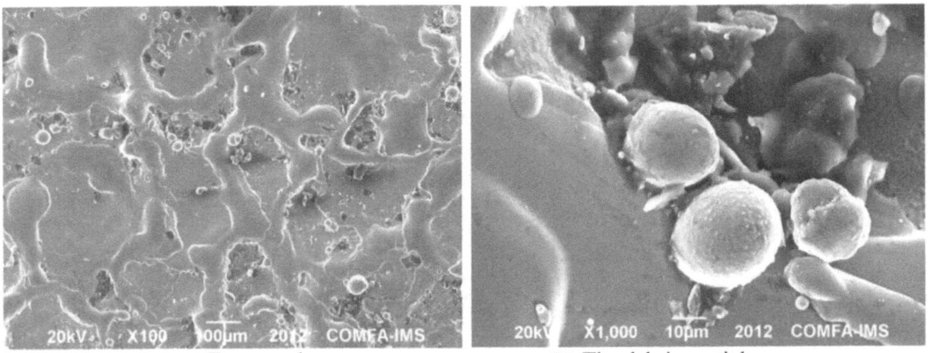

(a) Topography (b) The debris particles

Fig. 3.11: Machined surface after EDM (a) topography, (b) debris particles [96] (with kind permission from Springer Nature).

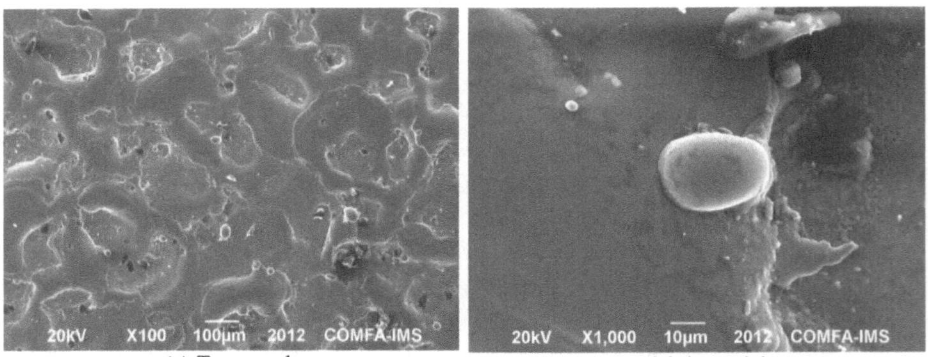

(a) Topography (b) The debris particles

Fig. 3.12: Machined surface after workpiece vibration in VAEDM (a) topography, (b) debris particles [96] (with kind permission from Springer Nature).

Despite its many advantages over conventional EDM, VAEDM still has several limitations. One notable issue is the nonuniformity in hole dimensions when machining arrays of microholes; central holes often become slightly larger than peripheral ones due to uneven debris distribution within the IEG [104]. Additionally, the process is highly sensitive to parameter tuning, as performance tends to decline beyond optimal

Tab. 3.3: Comparison between EDM and VAEDM.

Criteria	Electrodischarge machining (EDM)	Vibration-assisted electrodischarge machining (VAEDM)
Material removal rate [100, 101]	Lower, especially in µEDM and difficult-to-machine materials	Higher in both µEDM and Inconel 718 type materials
Surface quality and precision [92, 102]	Higher surface roughness and poorer circularity	Reduced surface roughness and improved hole circularity
Tool wear [100, 103]	Higher tool wear rate, larger taper and overcutting	Reduced tool wear rate, taper, and overcutting
Debris removal [101, 102]	Debris accumulation leads to short circuits and unstable discharges	Improved debris removal and charge stability; reduced molten material adherence and abnormal discharges.
Cost efficiency [101]	Higher per-part cost due to lower performance	Better performance and lower per-part cost compared to conventional EDM

settings, for instance, exceeding an ultrasonic amplitude of 5.22 µm resulted in reduced machining effectiveness [103]. While ultrasonic vibrations generally enhance surface quality, certain vibration conditions can actually lead to a slight increase in SR [101]. Moreover, the complexity of the system increases, requiring additional components and tighter control. The use of electrode arrays, for example, can restrict rotation, hinder debris removal, and further impact hole uniformity [104].

3.3.2.1 Latest Developments in Vibration-Assisted Electrodischarge Machining

Ultrasonic VAEDM significantly improves upon conventional EDM by augmenting debris evacuation and discharge stability for improved machining of advanced materials [105]. Its future lies in hybrid approaches and interdisciplinary advancements to meet growing demands for precision, efficiency, and sustainability. Integrating ultrasonic vibration with gas dielectrics like oxygen or argon boosts material removal while reducing environmental impact [100], and combining it with magnetic fields enhances debris evacuation via Lorentz forces. Research into nanofluid and biodegradable dielectrics further supports eco-friendly, high-performance machining [94]. Zhang et al. [90] introduced longitudinal-torsional ultrasonic vibration (LTV) electrodes that combine rotation with ultrasonic vibrations to enhance EDM drilling. Compared to conventional and longitudinal ultrasonic vibration electrodes, LTV electrodes significantly improve debris removal via higher flow velocities and turbulence. Computational fluid dynamics simulations and experiments showed a 1.5× increase in MRR, reduced TW, and the formation of positively tapered microholes suitable for injector applications (Fig. 3.13), marking a key advancement

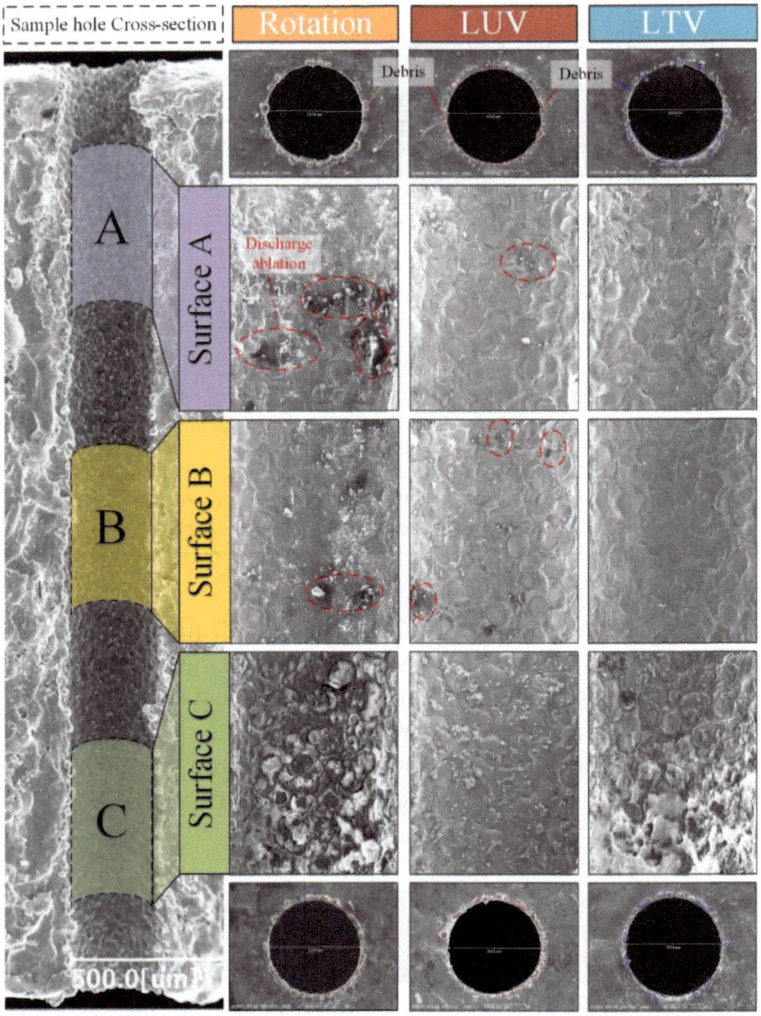

Fig. 3.13: Morphological features of microhole inlets, outlets, and inner walls under different machining strategies. The top row shows the inlet morphology, followed by Surface A representing the inner wall region near the inlet, Surface B corresponding to the central section of the inner wall, and Surface C indicating the region near the outlet. The bottom row illustrates the outlet morphology of the microholes [90] (with kind permission from Elsevier).

in VAEDM. Chen et al. [106] introduced a horizontal ultrasonic vibration system applied to the workpiece. This configuration improved discharge stability, enhanced machining efficiency by up to 42%, reduced TW from 5.6 to 3.6 mm, and significantly improved inner wall surface quality. Expanding the scope to cylindrical surfaces, Dinh et al. [107] designed a vibration head and electrode holder to deliver 20 kHz ultrasonic energy through a stationary hollow cylindrical copper electrode. By combining theoretical modeling with experimental validation, the study identified an optimal electrode holder length that enhanced MRR by as much as 35.5% relative to conventional EDM, without significantly affecting SR.

3.3.3 Magnetic Field-Assisted Electrodischarge Machining (MFAEDM)

In recent years, integrating magnetic fields into the EDM process, referred to as MFAEDM, has evolved as a promising development to further enhance machining efficiency and surface integrity [108]. The application of a magnetic field during EDM influences the plasma channel behavior and debris removal, thereby improving machining stability and overall performance. Magnetic fields have long been utilized in various nontraditional manufacturing techniques, including magnetic abrasive finishing and magnetic abrasive flow finishing, both recognized for their ability to achieve superior surface finishes on metals, ceramics, and composites [109]. Similarly, in MFAEDM, applying a magnetic field has been found to significantly improve key machining parameters like MRR, TWR, and surface finish, over those achieved with traditional EDM.

MFAEDM enhances the conventional EDM process by introducing an external magnetic field that interacts with the plasma channel formed during spark discharge [110]. The basic setup involves two electrodes, tool as cathode and workpiece as anode, both powered by an external DC pulse power source [109]. These components must be electrically conductive and are submerged in a dielectric fluid, typically hydrocarbon or mineral oil based, which serves multiple functions including spark stabilization, cooling, and debris flushing. A schematic representation of the MFAEDM setup is displayed in Fig. 3.14. As illustrated, magnetic poles are positioned on either side of the tool-workpiece interface, creating a magnetic field across the machining zone. The dielectric fluid is circulated using a pump, while a servo-controlled rotatable clamp ensures precise positioning of the tool. The applied magnetic field influences the motion of electrons and ions within the IEG, enhancing plasma channel stability and improving debris evacuation.

The integration of a magnetic field significantly alters the dynamics of the EDM process. The magnetic field interacts with the ionized plasma channel and charged particles within the IEG, forcing them to align along the magnetic lines of force [111–114]. This Lorentz force-induced motion enhances debris evacuation by disrupting particle accumulation, thereby reducing IEG and the occurrence of abnormal discharges. Figure 3.15 shows the interaction of the electric and magnetic fields generating a Lorentz force that alters the trajectory of debris particles. This force drives debris away from the IEG, improving machining efficiency, including reduced TW, higher MRR, and better process stability [115].

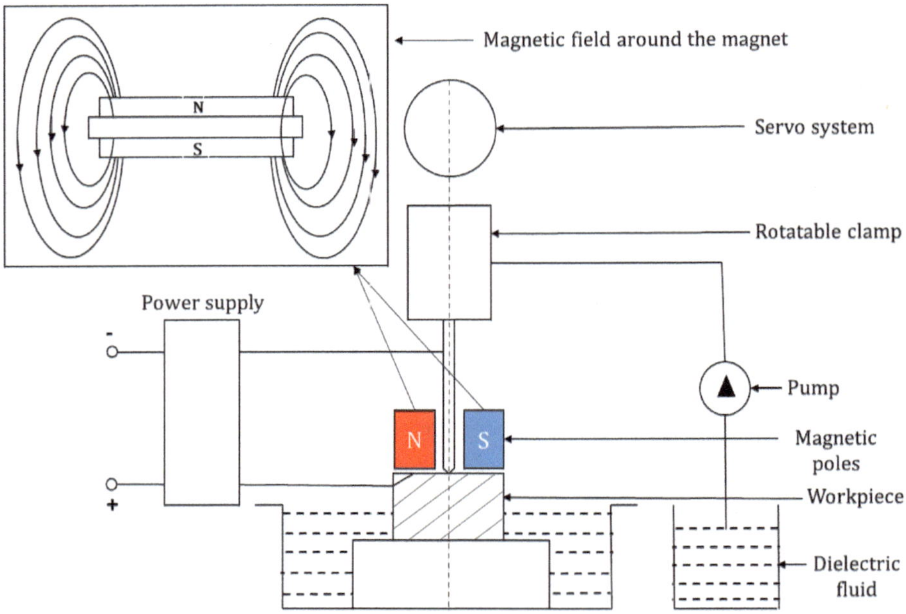

Fig. 3.14: Setup of magnetic field-assisted electrodischarge machining (MFAEDM).

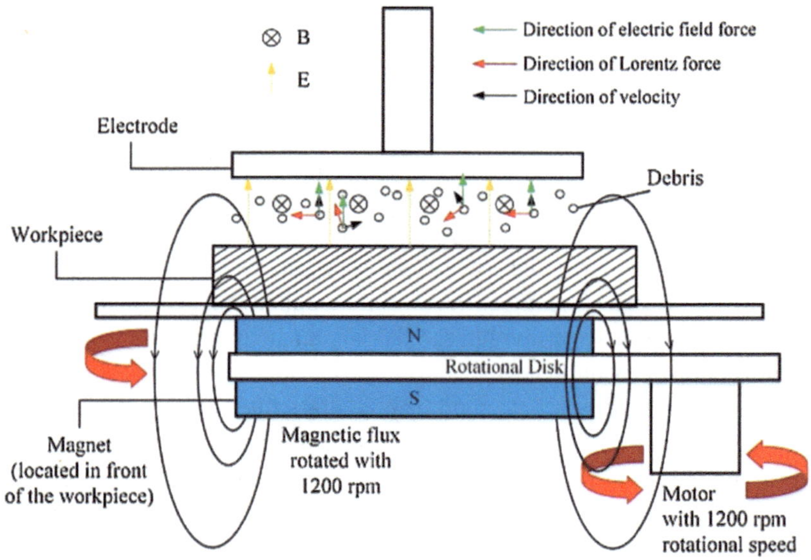

Fig. 3.15: Schematic diagram showing the movement of debris within the IEG under the influence of magnetic force [40, 115] (with kind permission from Elsevier).

The application of magnetic fields in various EDM processes has shown promising improvements in machining performance. In WEDM, introducing a constant magnetic field during finish cutting of ferromagnetic materials helps suppress divergence of electrons and shortens the IEG breakdown time, as supported by both theoretical modeling and single-pulse discharge experiments [116]. The magnetic field also enhances debris removal, especially when the Lorentz force aligns with dielectric flushing, leading to improved MRR. In μEDM, magnetic field assistance leverages magnetic forces and electrode rotation to stabilize discharges and enable deeper microhole machining [117]. For nonmagnetic materials, a parallel magnetic field induces directional currents and Ampere forces that increase plasma pressure and improve ablation efficiency. Though it significantly enhances machining precision and stability with a simple, low-cost setup, challenges like TW and tool deformation remain [118]. Magnetic field-assisted powder-mixed EDM combines a magnetic field with a powder-suspended dielectric to align conductive particles in the spark zone, increasing spark frequency and erosion rate. This leads to increased MRR and reduced SR, making it valuable for applications in toolmaking, aerospace, automotive, and biomedical sectors [119].

Magnetic field assistance in EDM offers several key advantages that improve machining performance. One of the most notable benefits is the increase in MRR. While the impact of magnetic field strength on MRR is moderate, studies show that higher field strengths can significantly enhance MRR, up to three times greater than conventional EDM under optimal conditions [40, 115]. This improvement is primarily due to better debris removal and enhanced plasma stability. Interestingly, while MRR increases with magnetic field strength, it may decline at higher I_p. In terms of discharge behavior, introducing an external magnetic field reduces arc pulses and promotes a more stable discharge waveform [120]. Enhanced debris evacuation reduces IEG, leading to a narrower plasma channel and higher energy density in the discharge zone. This contributes to improved machining precision and reduced abnormal discharges. In terms of surface finish, magnetic field assistance refines the spark generation process, producing more uniform and shallower craters. The concentrated and dense plasma channel enables controlled material erosion, reducing SR and improving overall finish quality [110, 120]. In some cases, SR has been reduced by up to one-third compared to traditional EDM, while also promoting beneficial alloying on the workpiece surface. Tab. 3.4 presents a comparison of key performance parameters between conventional EDM and MFAEDM, highlighting the latter's advantages in machining performance and surface quality. The cross-sectional comparison shown in Fig. 3.16 highlights the improved surface integrity achieved with MFAEDM [121]. Compared to conventional EDM, the magnetic-assisted process produces a thinner recast layer and fewer, shallower surface cracks due to more efficient debris expulsion and stable discharge conditions. This phenomenon is further visible in Fig. 3.17, where MFAEDM exhibits smoother morphology and fewer surface defects compared to conventional EDM [115]. These improvements result in better surface quality and reduced thermal damage to the workpiece material.

Tab. 3.4: Comparison between EDM and MFAEDM.

Criteria	Electrodischarge machining (EDM)	Magnetic field-assisted electrodischarge machining (MFAEDM)
Discharge status [115, 121]	High-intensity pulses with possible arcing	Controlled pulses with reduced arcing tendency
Material removal rate [115, 121–123]	Lower	Higher
Surface roughness [115, 121–124]	Higher	Lower
Tool wear rate [115, 123]	Lower	Higher
Recast layer thickness [115, 121]	Higher	Lower
Surface integrity [115, 121, 122, 124]	Longer and wider cracks, deeper crater	Shorter and narrower cracks, shallow crater

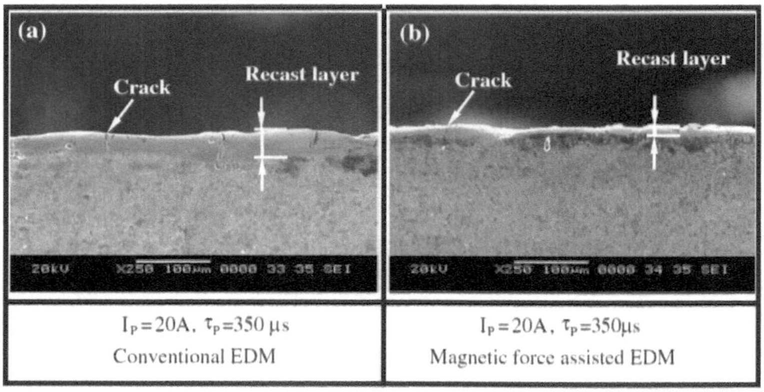

Fig. 3.16: Cross-sectional view of the surface machined using (a) conventional EDM, (b) magnetic-force-assisted EDM [121] (with kind permission from Springer Nature).

3.3.3.1 Latest Developments in Magnetic Field-Assisted Electrodischarge Machining

Recent advancements in MFAEDM focus on enhancing machining efficiency, surface integrity, and sustainability. In MFAEDM, mixing conductive powders into the dielectric enhances spark stability, debris removal, and MRR due to Lorentz force effects, while also improving surface hardness through particle deposition [110, 119]. This hybrid approach significantly boosts machining efficiency and surface integrity. Beravala and Pandey [108] examined MFAEDM using air and argon as dielectrics, report-

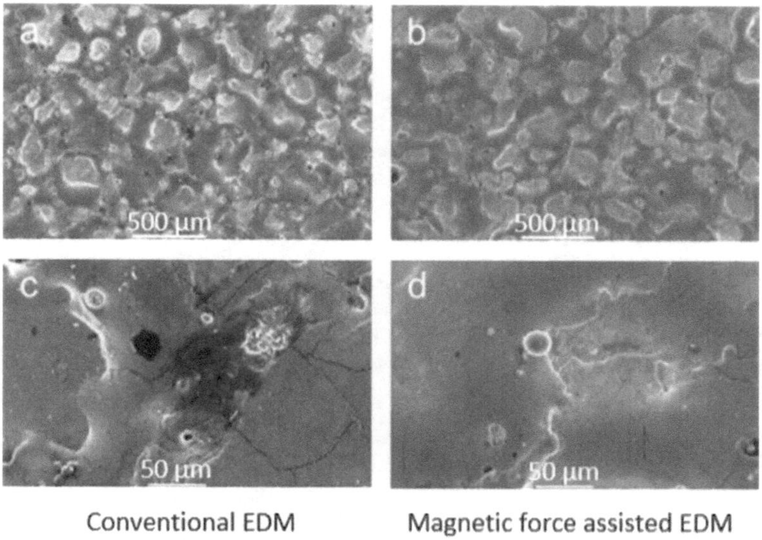

Fig. 3.17: Comparison of surfaces machined at $I_p = 20$ A, $T_{on} = 350$ µs using (a, c) conventional EDM, (b, d) magnetic force-assisted EDM [115] (with kind permission from Elsevier).

ing a 21–41% increase in MRR and up to a 16% reduction in TWR. Air facilitated exothermic reactions enhance MRR, while argon minimizes EWR due to its inert nature. Singh et al. [125] introduced a hybrid µEDM setup combining an orbital rotating magnetic field with ultrasonic vibrations, achieving improved debris removal, discharge stability, and reduced taper without compromising tool integrity. Similarly, Dong et al. [126] developed magnetic field-assisted ultrasonic EDM, which improved MRR, reduced SR by 32.6%, and decreased carbon build-up by 31.1%, thereby enhancing machining performance. Wang et al. [127] proposed an EDM milling technique enhanced by a rotating transverse magnetic field, which outperformed conventional and fixed-field EDM by producing a more uniform recast layer, fewer cracks, and improved corrosion resistance in sintered NdFeB, also demonstrating superior electrochemical properties. Zhang et al. [128] applied MFAEDM to high-volume SiCp/Al composites, achieving better debris evacuation, plasma stability, and a 6.6% increase in MRR alongside a 15% reduction in emissions, as optimized using a BPNN-GAQPSO model. Chu et al. [129] developed an EDM drilling approach utilizing adaptive planetary motion combined with a supporting magnetic field, achieving a 73.47% reduction in recast layer (Fig. 3.18) and an 18.52% improvement in efficiency by dynamically adjusting planetary motion with a rotating magnetic field.

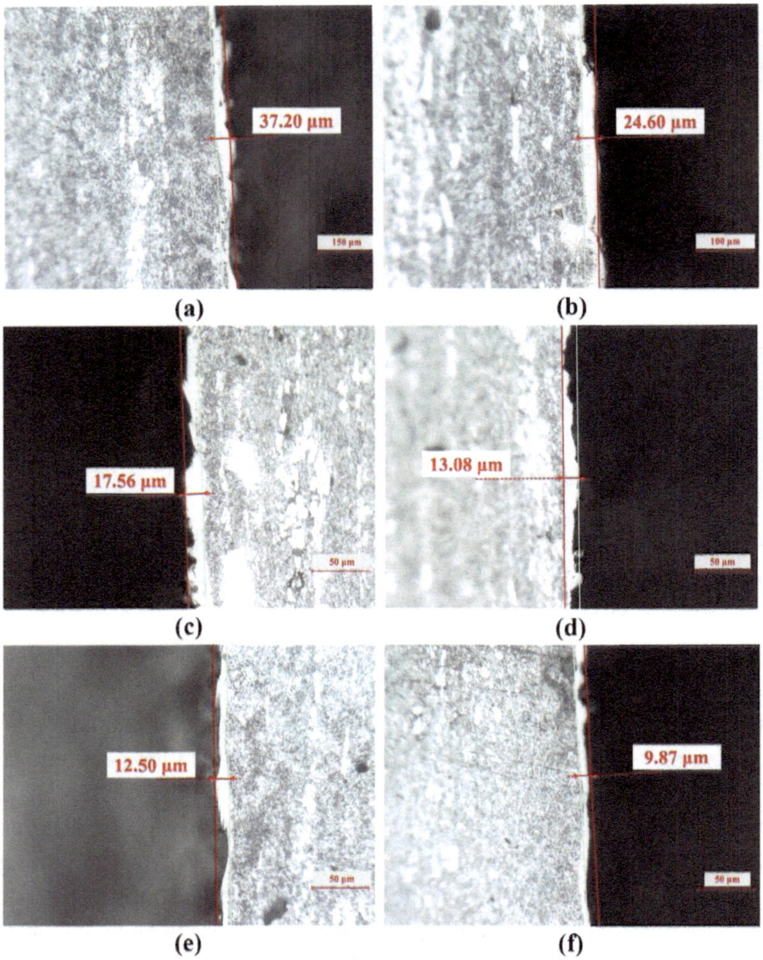

Fig. 3.18: Recast layer thickness under various machining conditions: (a) conventional EDM, (b) EDM with an auxiliary magnetic field, (c) EDM using planetary tool motion, (d) planetary motion combined with a magnetic field, (e) adaptive planetary control, and (f) adaptive planetary control with magnetic field assistance [129] (with kind permission from Springer Nature).

3.4 Conclusion and Future Directions

EDM, while revolutionary in machining hard and intricate materials, still presents limitations like low MRR, high TW, and surface defects. Hybrid EDM processes have successfully addressed many of these issues by incorporating complementary techniques. As featured in this chapter, EDG, EDAG, and EDDG utilize rotating conductive tools and abrasives to enhance machining efficiency and surface finish. LAEDM improves precision and discharge efficiency through localized preheating, while VAEDM boosts debris removal and discharge stability. MFAEDM introduces Lorentz forces that refine plasma behavior and debris evacuation. These hybrid approaches have collectively broadened EDM's scope, enabling more efficient, precise, and application-specific machining across sectors like aerospace, biomedical, and micromanufacturing.

As EDM technology advances, hybrid systems are poised to play an increasingly transformative role. Future EDM platforms will integrate in-situ microtool fabrication with machining operations to enhance precision and eliminate setup errors in micromachining. Emerging technologies like magnetic levitation EDM, which employ magnetic levitation-based servo systems, offer faster and more stable IEG control, thereby improving machining quality and reducing defects. In EDG, the use of metal-bonded diamond wheels promises higher MRR and improved dimensional accuracy. Enhancing bond matrix conductivity and grit retention can significantly boost tool performance. Furthermore, adding conductive powders to the dielectric can enlarge the IEG, resulting in better surface finishes. EDG systems can also be augmented with ultrasonic vibrations, magnetic field assistance, or cryogenic cooling to enhance debris removal and reduce TW. For LAEDM, picosecond laser pulses show strong potential for minimizing HAZ and improving surface finish, while Ytterbium pulsed fiber lasers can reduce taper angles and thermal damage. The combination of EDM with SLM or 3D printing can enable the production of intricate components with high durability and superior surface quality. Ultrasonic VAEDM, when paired with gas-based dielectrics like oxygen or argon, can significantly increase MRR and support more sustainable machining. Furthermore, the development of biodegradable dielectrics aligns well with green manufacturing goals. Multidirectional vibration systems are also emerging to enable deeper and more uniform material removal. Further enhancements in MFAEDM can be achieved by combining it with powder-mixed dielectrics to improve conductivity and spark distribution. The addition of ultrasonic vibrations helps stabilize discharges, while rotary planetary motion introduces centrifugal action for superior debris evacuation. Finally, intelligent control is set to redefine EDM's capabilities. The incorporation of advanced optimization techniques like genetic algorithm, particle swarm optimization, teacher learning-based optimization, and continuous ant colony optimization can fine-tune parameters for maximum efficiency and surface quality. Coupling these systems with real-time sensor data and artificial intelligence-driven feedback control will enable adaptive, self-optimizing EDM platforms tailored for Industry 4.0.

Nomenclature

EDM	Electrodischarge machining
IEG	Interelectrode gap
WEDM	Wire electrodischarge machining
MRR	Material removal rate

T_{on}	Pulse duration
DE	Discharge energy
TW	Tool wear
SR	Surface roughness
I_p	Current
T_{off}	Pulse interval
HAZ	Heat affected zone
EDAG	Electrodischarge abrasive grinding
VAEDM	Vibration-assisted EDM
LAEDM	Laser-assisted EDM
MFAEDM	Magnetic field-assisted EDM
EDG	Electrodischarge grinding
DC	Direct current
DI	Deionized
TWR	Tool wear rate
EDDG	Electrodischarge diamond grinding
SEM	Scanning electron microscopy
μEDM	Micro electrodischarge machining
MT	Machining time
LBMM	Laser beam micromachining
SLM	Selective laser melting
LTV	Longitudinal-torsional ultrasonic vibration
LUV	Longitudinal ultrasonic vibration

References

[1] Rajurkar KP, Wang WM. Improvement of EDM performance with advanced monitoring and control systems, J Manuf Sci Eng. 1997;119(4B):770–775.

[2] Ho KH, Newman ST. State of the art electrical discharge machining (EDM), Int J Mach Tools Manuf. 2003;43(13):1287–1300.

[3] Hassan E-H. Advanced machining processes: Nontraditional and hybrid machining processes. New York: McGraw-Hill Professional Publishing; 2005.

[4] Jameson EC. Electrical discharge machining: Tooling, methods, and applications, Society of Manufacturing Engineers. 1983; [cited 2025 Oct 31]. Available from: [Online].

[5] Nafi MA, Jahan MP. Functional surface generation by EDM – a review, Micromachines. 2022;14(1):115.

[6] Ahmed A, Fardin A, Tanjilul M. et al. A comparative study on the modelling of EDM and hybrid electrical discharge and arc machining considering latent heat and temperature-dependent properties of Inconel 718, Int J Adv Manuf Technol. 2018;94(5–8):2729–2737.

[7] Mohd Abbas N, Solomon DG, Fuad Bahari M. A review on current research trends in electrical discharge machining (EDM), Int J Mach Tools Manuf. 2007;47(7–8):1214–1228.

[8] Xie B, Hou J, Dong B. et al. Thermal model of crater formation process in electrical discharge machining, Case Stud Therm Eng. 2024;56:104303.

[9] Gadalla AM, Petrofes NF. Surfaces of advanced ceramic composites formed by electrical discharge machining, Mater Manuf Process. 1990;5(2):253–271.

[10] Ojha N, Zeller F, Mueller C. et al. The effect of the electrical discharge machining process on the material properties of nonconductive ceramics, J Micro Nano-Manuf. 2016;4(1):1–5.

[11] Sultan T, Kumar A, Gupta RD. Material removal rate, electrode wear rate, and surface roughness evaluation in die sinking EDM with hollow tool through response surface methodology, Int J Manuf Eng. 2014;2014:1–16.
[12] Gostimirovic M, Kovac P, Sekulic M. et al. Influence of discharge energy on machining characteristics in EDM, J Mech Sci Technol. 2012;26(1):173–179.
[13] Torres A, Puertas I, Luis CJ. Modelling of surface finish, electrode wear and material removal rate in electrical discharge machining of hard-to-machine alloys, Precis Eng. 2015;40:33–45.
[14] Hegde A, Shetty R, Nayak R. et al. Optimizing the die-sink EDM machinability of AISI 316L using Ti-6Al-4V-SiCp electrodes: a computational approach, J Manuf Mater Process. 2024;8(5):202.
[15] Qudeiri JEA, Zaiout A, Mourad A-HI. et al. Principles and characteristics of different EDM processes in machining tool and die steels, Appl Sci. 2020;10(6):2082.
[16] Liu S, Thangaraj M, Moiduddin K. et al. Influence of adaptive gap control mechanism and tool electrodes on machining titanium (Ti-6Al-4V) alloy in EDM process, Materials (Basel). 2022;15(2):513.
[17] Xin B, Gao M, Li S. et al. Modeling of interelectrode gap in electric discharge machining and minimum variance self-tuning control of interelectrode gap, Math Probl Eng. 2020;2020:1–20.
[18] Sisodiya MS, Shukla S, Bajpai V. Feasibility analysis of novel Maglev EDM by comparing with conventional micro EDM, Sci Rep. 2022;12(1):2613.
[19] Dhakar K, Kumar R, Katheria A. et al. Effect of various dielectric fluids on electric discharge machining (EDM): A review, J Braz Soc Mech Sci Eng. 2022;44(10):1–17.
[20] Chakraborty S, Dey V, Ghosh SK. A review on the use of dielectric fluids and their effects in electrical discharge machining characteristics, Precis Eng. 2015;40:1–6.
[21] Ablyaz TR, Shlykov ES, Muratov KR. Modeling of EDM process flushing mechanism, Materials (Basel). 2023;16(11):4158.
[22] Jiang Y, Kong L, Ping X. et al. Utilizing a porous electrode for the flushing fluid in electrical discharge machining, J Manuf Process. 2021;62:248–256.
[23] Kar S, Patowari PK. Electrode wear phenomenon and its compensation in micro electrical discharge milling: a review, Mater Manuf Process. 2018;33(14):1491–1517.
[24] AG V, B S, BK L. et al. A decision-making model for tool material selection in electrical discharge machining, Mater Today Proc. 2022;59(3).
[25] Mahardika M, Tsujimoto T, Mitsui K. A new approach on the determination of ease of machining by EDM processes, Int J Mach Tools Manuf. 2008;48(7–8).
[26] Singh S, Maheshwari S, Pandey PC. Some investigations into the electric discharge machining of hardened tool steel using different electrode materials, J Mater Process Technol. 2004;149(1–3).
[27] Mao X, Almeida S, Mo J. et al. The state of the art of electrical discharge drilling: A review, Int J Adv Manuf Technol. 2022;121(5–6).
[28] Volosova M, Okunkova A, Peretyagin P. et al. On electrical discharge machining of non-conductive ceramics: A review, Technologies. 2019;7(3):1–18.
[29] Sarmah A, Kar S, Patowari PK. Surface alteration of Al-7075 alloy by depositing Inconel 718 through electrical discharge coating, J Mater Eng Perform. 2025;34(12):11765–11778.
[30] Altan T, Lilly B, Yen YC. et al. Manufacturing of dies and molds, CIRP Ann. 2001;50(2):404–422.
[31] Ablyaz TR, Shlykov ES, Muratov KR. et al. Study of wire-cut electro-discharge machining of heat-resistant nickel alloys, Materials (Basel). 2023;16(20):1–15.
[32] Büssemaker H, Meinshausen A-K, Bui VD. et al. Silver-integrated EDM processing of TiAl6V4 implant material has antibacterial capacity while optimizing osseointegration, Bioact Mater. 2024;31:497–508.
[33] Chaitanya CRA, Wang N, Takahata K. MEMS-based micro-electro-discharge machining (M3EDM) by electrostatic actuation of machining electrodes on the workpiece, J Microelectromech Syst. 2010;19 (3):690–699.

[34] Rajagopal SP, Ganesh V, Lanjewar AV. et al. Past and current status of hybrid electric discharge machining (H-EDM) processes, Int J Adv Mater Manuf Charact. 2013;3(1):111–118.
[35] Shrivastava PK, Dubey AK. Electrical discharge machining–based hybrid machining processes: A review, Proc Inst Mech Eng Part B J Eng Manuf. 2014;228(6):799–825.
[36] Jain VK, Mote RG. On the temperature and specific energy during electrodischarge diamond grinding (EDDG), Int J Adv Manuf Technol. 2005;26(1–2).
[37] Wüthrich R, Fascio V. Machining of non-conducting materials using electrochemical discharge phenomenon – an overview, Int J Mach Tools Manuf. 2005;45(9):1095–1108.
[38] Zhixin J, Jianhua Z, Xing A. Study on a new kind of combined machining technology of ultrasonic machining and electrical discharge machining, Int J Mach Tools Manuf. 1997;37(2):193–199.
[39] Liang HY, Kuo CL, Huang JD. Precise micro-assembly through an integration of micro- EDM and Nd-YAG, Int J Adv Manuf Technol. 2002;20(6):454–458.
[40] Zhang Z, Zhang Y, Ming W. et al. A review on magnetic field assisted electrical discharge machining, J Manuf Process. 2021;64:694–722.
[41] Zeng Z, Wang Y, Wang Z. et al. A study of micro-EDM and micro-ECM combined milling for 3D metallic micro-structures, Precis Eng. 2012;36(3):500–509.
[42] Ji R, Liu Y, Zhang Y. et al. Electric discharge milling and mechanical grinding compound machining of silicon carbide ceramic, Mach Sci Technol. 2010;14(4):502–521.
[43] Shih HR, Shu KM. A study of electrical discharge grinding using a rotary disk electrode, Int J Adv Manuf Technol. 2008;38(1–2).
[44] Nair AR, Paul L. Advancements in abrasive electrical discharge grinding (AEDG): A review, Mater Today Proc. 2023;72:2897–2903.
[45] Gupta K, Jain NK, Laubscher RF. Hybrid machining processes. Cham: Springer International Publishing; 2016. [cited 2025 Oct 31]. Available from: [Online].
[46] Lu Y, Mou X, Xiao H. et al. Surface integrity of binderless WC using dry electrical discharge assisted grinding, Int J Precis Eng Manuf Green Technol. 2024;12(1):43–60.
[47] Kulshrestha AS, Mertiya AS, Unune DR. et al. A review on electrical discharge grinding: current status and future perspectives, Surf Rev Lett. 2023;30(3):1–25.
[48] Kulshrestha AS, Unune DR, Dargar AK. Investigation and optimization of electro discharge face grinding of Monel 400 alloy using response surface methodology and genetic algorithm, Sadhana. 2023;48(3):1–17.
[49] Kumar Singh Yadav S, Yadava V, Lakshmi Narayana V. Experimental study and parameter design of electro-discharge diamond grinding, Int J Adv Manuf Technol. 2008;36(1–2):34–42.
[50] Satyarthi MK, Pandey PM. Modeling of material removal rate in electric discharge grinding process, Int J Mach Tools Manuf. 2013;74:65–73.
[51] Lu Y, Huang Y, Liu X. et al. Study on dry electrical discharge assisted grinding of SiCp/Al composite, Chin J Mech Eng. 2025;38(1):11.
[52] Koshy P, Jain VK, Lal GK. Mechanism of material removal in electrical discharge diamond grinding, Int J Mach Tools Manuf. 1996;36(10):1173–1185.
[53] Strelchuk R, Shelkovyi O. Research of the cutting mechanism at electrical discharge grinding, Cut Tools Technol Syst. 2021;1(95):37–44.
[54] Kozak J, Zybura-Skrabalak M, Skrabalak G. Development of advanced abrasive electrical discharge grinding (AEDG) system for machining difficult-to-cut materials, Procedia CIRP. 2016;42:872–877.
[55] Ding W, Li H, Zhang L. et al. Diamond wheel dressing: A comprehensive review, J Manuf Sci Eng. 2017;139(12):1–27.
[56] Kishore K, Sinha MK, Singh A. et al. A comprehensive review on the grinding process: Advancements, applications and challenges, Proc Inst Mech Eng Part C J Mech Eng Sci. 2022;236 (22):10923–10952.

[57] Satyarthi MK, Pandey PM. Surface and subsurface study on EDG and EDM processing of advanced ceramic composite material, In ASME International Mechanical Engineering Congress and Exposition (IMECE). American Society of Mechanical Engineers; 2013. [cited 2025 Oct 31].

[58] Patel KM, Pandey PM, Venkateswara Rao P. Surface integrity and material removal mechanisms associated with the EDM of Al_2O_3 ceramic composite, Int J Refract Met Hard Mater. 2009;27 (5):892–899.

[59] Satyarthi MK, Pandey PM. Comparison of EDG, diamond grinding, and EDM processing of conductive alumina ceramic composite, Mater Manuf Process. 2013;28(4):369–374.

[60] RS Y, Yadava V. Experimental investigations on electrical discharge diamond face surface grinding (EDDFSG) of hybrid metal matrix composite, Mater Manuf Process. 2017;32(2):135–144.

[61] Singh GK, Yadava V, Kumar R. Diamond face grinding of WC-Co composite with spark assistance: Experimental study and parameter optimization, Int J Precis Eng Manuf. 2010;11(4):509–518.

[62] RN Y, Yadava V. Slotted-electrical discharge diamond cut-off grinding of Al/SiC/B_4C hybrid metal matrix composite, J Mech Sci Technol. 2014;28(1):309–316.

[63] Masuzawa T, Fujino M, Kobayashi K. et al. Wire electro-discharge grinding for micro- machining, CIRP Ann Manuf Technol. 1985;34(1):431–434.

[64] Zhang L, Tong H, Li Y. Precision machining of micro tool electrodes in micro EDM for drilling array micro holes, Precis Eng. 2015;39:100–106.

[65] Kumar M, Satsangi PS. A study on machining performance of wire electric discharge grinding (WEDG) process during machining of tungsten alloy micro-tools, Sadhana. 2021;46(2):1–12.

[66] Parthiban M, Krishnaraj V, Sindhumathi R. et al. Investigation on manufacturing of microtools made of tungsten carbide using wire electric discharge grinding (WEDG), J Braz Soc Mech Sci Eng. 2017;39 (9):3571–3580.

[67] Zhao WS, Meng QG, Wang ZL. The application of research on powder mixed EDM in rough machining, J Mater Process Technol. 2002;129(1–3).

[68] Satyarthi MK, Pandey PM. Processing of Al_2O_3–SiCw–TiC ceramic composite by powder mixed electric discharge grinding, Mech Adv Mater Mod Process. 2016;2(1):5.

[69] Unune DR, Marani Barzani M, Mohite SS. et al. Fuzzy logic-based model for predicting material removal rate and average surface roughness of machined Nimonic 80A using abrasive-mixed electro-discharge diamond surface grinding, Neural Comput Appl. 2018;29(9):647–662.

[70] Jahan MP. Electro-discharge machining (EDM): Principle of the EDM process, Modern Manufacturing Processes. 2019;377–409.

[71] Huang JD, Kuo CL. Pin-plate micro assembly by integrating micro-EDM and Nd-YAG laser, Int J Mach Tools Manuf. 2002;42(13):1455–1464.

[72] Kuo CL, Huang JD, Liang HY. et al. Fabrication of 3D metal microstructures using a hybrid process of micro-EDM and laser assembly, Int J Adv Manuf Technol. 2003;21(10–11).

[73] Li L, Diver C, Atkinson J. et al. Sequential laser and EDM micro-drilling for next generation fuel injection nozzle manufacture, CIRP Ann. 2006;55(1):179–182.

[74] Weber P, Haupt S, Schulze V. Hybrid machining of microstructures using a combination of electrical discharge machining milling and laser ablation, J Vac Sci Technol B. 2009;27(3):1327–1329.

[75] Lee CM, Woo WS, Kim DH. et al. Laser-assisted hybrid processes: A review, Int J Precis Eng Manuf. 2016;17(2):257–267.

[76] Clerici M, Hu Y, Lassonde P. et al. Laser-assisted guiding of electric discharges around objects, Sci Adv. 2015;1(5):1–6.

[77] Al Javed MO, Bin Rashid A. Laser-assisted micromachining techniques: An overview of principles, processes, and applications, Adv Mater Process Technol. 2024;0(0):1–44.

[78] Rashid MA, Rahman M, Senthil Kumar A. A study on compound micromachining using laser and electric discharge machining (EDM), Adv Mater Process Technol. 2016;2(2):258–265.

[79] Rashid MAN, Saleh T, Noor WI. et al. Effect of laser parameters on sequential laser beam micromachining and micro electro-discharge machining, Int J Adv Manuf Technol. 2021;114(3–4).
[80] Al-Ahmari AMA, Rasheed MS, Mohammed MK. et al. A hybrid machining process combining micro-EDM and laser beam machining of nickel-titanium-based shape memory alloy, Mater Manuf Process. 2016;31(4):447–455.
[81] Rashid MAN, Saleh T, Abdul Hamid SB. et al. Analysis and modelling of laser-micro EDM-based hybrid micro milling on stainless steel (SUS304) using Box-Behnken design, IIUM Eng Congr Proc. 2023;1(1):1–10.
[82] Sahu AK, Mahapatra SS. Performance analysis of tool electrode prepared through laser sintering process during electrical discharge machining of titanium, Int J Adv Manuf Technol. 2020;106(3–4).
[83] Shiu PP, Knopf GK, Ostojic M. Fabrication of metallic micromolds by laser and electro- discharge micromachining, Microsyst Technol. 2010;16(3):477–485.
[84] Wälder G, Richard J. Removal of the heat affect zone created by EDM with pico-second laser machining, Procedia CIRP. 2016;42:475–480.
[85] Hassanin H, Modica F, El-Sayed MA. et al. Manufacturing of Ti–6Al–4V micro- implantable parts using hybrid selective laser melting and micro-electrical discharge machining, Adv Eng Mater. 2016;18(9):1544–1549.
[86] Chung DK, Han CH, Choi YJ. et al. Laser EDM hybrid micro machining of CFRP, J Korean Soc Precis Eng. 2023;40(2):99–104.
[87] Song KY, Kim GH, Shin J. A hybrid manufacturing process for a microgripper using selective laser melting 3D printing and wire EDM, J Mech Sci Technol. 2023;37(4):1931–1937.
[88] Ma Y, Liang Z, Du Y. et al. Fabrication method of polycrystalline diamond micro-drill using a hybrid process of pulsed laser, EDM and precision grinding, Int J Refract Met Hard Mater. 2025;129:107124.
[89] Wang Y, Fan L, Shi J. et al. Effect of cavitation on surface formation mechanism of ultrasonic vibration-assisted EDM, Int J Adv Manuf Technol. 2023;124(10):3645–3656.
[90] Zhang P, Yin Z, Yu M. et al. Investigating mechanisms of debris removal in ultrasonic vibration-assisted EDM drilling, Int J Mech Sci. 2024;279:109486.
[91] Kremer D, Lebrun JL, Hosari B. et al. Effects of ultrasonic vibrations on the performances in EDM, CIRP Ann Manuf Technol. 1989;38(1):199–202.
[92] Sharma AK, Singh V, Goyal A. et al. Experimental analysis of Inconel 625 alloy to enhance the dimensional accuracy with vibration-assisted micro-EDM, Int J Interact Des Manuf. 2024;18 (10):7495–7509.
[93] Unune DR, Mali HS. Current status and applications of hybrid micro-machining processes: A review, Proc Inst Mech Eng Part B J Eng Manuf. 2015;229(10):1681–1693.
[94] Prihandana GS, Mahardika M, Hamdi M. et al. Effect of micro-powder suspension and ultrasonic vibration of dielectric fluid in micro-EDM processes: Taguchi approach, Int J Mach Tools Manuf. 2009;49(12–13).
[95] Gao C, Liu Z. A study of ultrasonically aided micro-electrical-discharge machining by the application of workpiece vibration, J Mater Process Technol. 2003;139(1–3):226–228.
[96] Huu PN, Tien LB, Duc QT. et al. Multi-objective optimization of process parameter in EDM using low-frequency vibration of workpiece assigned for SKD61, Sādhanā. 2019;44(10):211.
[97] Tsai MY, Fang CS, Yen MH. Vibration-assisted electrical discharge machining of grooves in a titanium alloy (Ti-6Al-4V), Int J Adv Manuf Technol. 2018;97(1–4):297–304.
[98] Gupta K, Sarmah P, Pawanr S. Analysis of recent research and innovations in vibration- assisted hybrid machining processes, J Eng Technol Sci. 2025;57(2):273–288.
[99] Zhang QH, Du R, Zhang JH. et al. An investigation of ultrasonic-assisted electrical discharge machining in gas, Int J Mach Tools Manuf. 2006;46(12–13):1582–1588.
[100] Xing Q, Yao Z, Zhang Q. Effects of processing parameters on processing performances of ultrasonic vibration-assisted micro-EDM, Int J Adv Manuf Technol. 2021;112(1–2):71–86.

[101] Panda D, Sahu SK, Das SR. Comparative performance evaluation and economic assessment between traditional and vibration-assisted EDM during machining of Inconel 718, Proc Inst Mech Eng Part C J Mech Eng Sci. 2024;238(16):8197–8219.
[102] Hou S, Bai J, Liu H. et al. Study on material erosion mechanism of ultrasonic vibration- assisted micro-EDM based on heat-flow coupling analysis, Int J Adv Manuf Technol. 2023;125(1–2).
[103] Zhang P, Yin Z, Dai C. et al. The effect of ultrasonic amplitude on the performance of ultrasonic vibration-assisted EDM micro-hole machining, Int J Adv Manuf Technol. 2022;122(3–4).
[104] Zhang Y, Xie B. Investigation on hole diameter non-uniformity of hole arrays by ultrasonic vibration-assisted EDM, Int J Adv Manuf Technol. 2021;112(11–12):3083–3091.
[105] Sabyrov N, Jahan MP, Bilal A. et al. Ultrasonic vibration-assisted electro-discharge machining (EDM) – an overview, Materials (Basel). 2019;12(3):522.
[106] Chen B, Tao M, Luo Z. Effect of ultrasonic-assisted EDM based on horizontal vibration on deep and small hole machining, Sci Rep. 2024;14(1):1–11.
[107] Dinh VT, Do TT, Le TQ. et al. Initial research on ultrasonic vibration-assisted EDM for processing cylindrical surfaces, Coatings. 2025;15(4):463.
[108] Beravala H, Pandey PM. Experimental investigations to evaluate the effect of magnetic field on the performance of air and argon gas assisted EDM processes, J Manuf Process. 2018;34:356–373.
[109] Teimouri R, Baseri H. Study of tool wear and overcut in EDM process with rotary tool and magnetic field, Adv Tribol. 2012;2012:1–8.
[110] Bhatt G, Batish A, Bhattacharya A. Experimental investigation of magnetic field- assisted powder mixed electric discharge machining, Part Sci Technol. 2015;33(3):246–256.
[111] Teimouri R, Baseri H. Optimization of magnetic field-assisted EDM using the continuous ACO algorithm, Appl Soft Comput J. 2014;14C:381–9.
[112] Zabihi SS, Etemadi Haghighi S, Soleimanimehr H. et al. Effects of auxiliary magnetic field strength and direction on material removal rate and surface roughness in magnetic field-assisted electrical discharge machining, CIRP J Manuf Sci Technol. 2023;41:446–452.
[113] Bains PS, Sidhu SS, Payal HS. Magnetic field-assisted EDM: New horizons for improved surface properties, Silicon. 2018;10(4):1275–1282.
[114] Shabgard MR, Gholipoor A, Mohammadpourfard M. Investigating the effects of external magnetic field on machining characteristics of electrical discharge machining process, numerically and experimentally, Int J Adv Manuf Technol. 2019;102(1–4):55–65.
[115] Lin YC, Lee HS. Machining characteristics of magnetic force-assisted EDM, Int J Mach Tools Manuf. 2008;48(11):1179–1186.
[116] Chen Z, Zhang Y, Zhang G. et al. Theoretical and experimental study of magnetic- assisted finish cutting of ferromagnetic material in WEDM, Int J Mach Tools Manuf. 2017;123:36–47.
[117] Yeo SH, Murali M, Cheah HT. Magnetic field assisted micro electro-discharge machining, J Micromech Microeng. 2004;14(11):1526–1529.
[118] Heinz K, Kapoor SG, DeVor RE. et al. An investigation of magnetic-field-assisted material removal in micro-EDM for nonmagnetic materials, J Manuf Sci Eng. 2011;133(2):021002.
[119] Rouniyar AK, Shandilya P. Optimization of process parameters in magnetic field assisted powder mixed EDM of aluminium 6061 alloy, Proc Inst Mech Eng Part C J Mech Eng Sci. 2021;235 (16):2998–3014.
[120] Renjith R, Paul L. A review on magnetic field assisted micro machining, IOP Conf Ser Mater Sci Eng. 2018;396(1):012074.
[121] Lin YC, Lee HS. Optimization of machining parameters using magnetic-force-assisted EDM based on grey relational analysis, Int J Adv Manuf Technol. 2009;42(11–12):1052–1064.
[122] Efendee AM, Azhari A, Hasnain AR. et al. Investigation of magnetic field effect on MRR, EWR and surface roughness during EDM of AISI 420 tool steel, IOP Conf Ser Mater Sci Eng. 2021;1068 (1):012018.

[123] Gholipoor A, Baseri H, Shakeri M. et al. Investigation of the effects of magnetic field on near-dry electrical discharge machining performance, Proc Inst Mech Eng Part B J Eng Manuf. 2016;230 (4):744–751.

[124] Govindan P, Gupta A, Joshi SS. et al. Single-spark analysis of removal phenomenon in magnetic field assisted dry EDM, J Mater Process Technol. 2013;213(7):1048–1058.

[125] Singh G, Satsangi PS, Prajapati DR. Effect of rotating magnetic field and ultrasonic vibration on micro-EDM process, Arab J Sci Eng. 2020;45(2):1059–1070.

[126] Dong Y, Zhang S, Wang Y. et al. Magnetic field-assisted ultrasonic EDM removal mechanism and surface morphology study, Int J Adv Manuf Technol. 2024;130(11–12):6043–6057.

[127] Wang W, Bai X, Yang TY. et al. Research on surface corrosion resistance of sintered NdFeB by rotating transverse magnetic field assisted EDM-milling, CIRP J Manuf Sci Technol. 2024;52:318–329.

[128] Zhang Z, Zhang Y, Lin L. et al. Study on productivity and aerosol emissions of magnetic field-assisted EDM process of SiCp/Al composite with high volume fractions, J Clean Prod. 2021;292:126018.

[129] Chu H, Li Z, Xi X. et al. Adaptive planetary control with auxiliary magnetic field method for fast EDM drilling, Int J Adv Manuf Technol. 2022;120(3–4).

Mahesh Kumar Sah, Harpreet Singh

4 Hybrid Chemo-magnetorheological Finishing for Optical Applications

Abstract: Hybrid finishing techniques have a direct impact on optical qualities, bearing efficiency, resilience to corrosion, suitability with biomedical applications, and microfluidics features. The surface quality of parts not only significantly affects their usability, service life, and reliability, but is also closely related to the functionality, longevity, and efficiency of the full product. Therefore, the market demand for nano-finished components is rising day by day, which has to be fulfilled by hybrid finishing techniques. Conversely, there is a growing trend in the use of aluminum alloy-based components in sectors such as aerospace, defense, optics, automobiles, electronics, and nuclear applications. Aluminum-based alloys are extensively utilized due to their lightweight nature and excellent corrosion resistance, which contribute to high performance; however, they exhibit significant ductility. Their flexible and malleable structure often leads to surface abrasions during processing, making the finishing stage more time-consuming and less precise. Consequently, visible marks are frequently left on the surface, which can diminish the performance and appearance of the final component surface. Recently, hybrid chemo-magnetorheological finishing (HCMRF) of aluminum materials for high-performance optical components has been studied to achieve the desired result. This chapter provides an overview of several optical applications employing aluminum and its alloys, as well as the influence of HCMRF on the finishing of aluminum alloys.

Keywords: HCMRF, aluminum alloys, processing parameter, surface roughness, optical applications

4.1 Introduction

The establishment of surfaces into products has become increasingly widespread in the industry sector, with the goal of improving both the functionality and aesthetics [1]. Optical lens surfaces play a crucial role in controlling light through mechanisms of convergence and divergence, making them indispensable across a broad spectrum of applications from corrective eyewear to high-precision scientific instruments such as microscopes and telescopes. The performance of these optical systems is intricately tied to the quality of the lens surfaces, which must exhibit precision nano-finished profile. Even minor surface imperfections can scatter light, lower image contrast, and compromise the overall effectiveness of the system. Therefore, the final finishing

 | https://doi.org/10.1515/9783111704456-004

stage, namely finishing, is a critical process in the manufacturing of optical components, as it directly determines their optical quality and functional reliability. However, the intricate geometries and the properties of aluminum-based alloys present unique challenges in achieving the desired results. The challenges have been surmounted by the formulation of several techniques, including refining the surface finishing process [2]. A satisfactory degree of surface smoothness can be achieved using conventional processes, namely grinding, lapping, chemical treatment, and abrasive blasting. Nevertheless, they are typically inadequate for accurately finishing intricate geometries due to restricted force control and the application of elevated pressures, which can compromise the surface and introduce faults. Conversely, advanced finishing techniques are employed to address the limitations of traditional methods, utilizing electrical, thermal, chemical, mechanical, or hybrid processes [3–5]. These sophisticated methods provide high-precision material removal, providing exceptional surface smoothness and the capability to machine complex forms without generating burrs or residual stresses [6, 7].

Considering the present challenges faced in the advanced finishing processes, the work reported in this chapter attempts to present a novel surface enhancement technique as a possible solution of finishing optical and other products. Hybrid chemo-magnetorheological finishing (HCMRF) diminishes average surface roughness (R_a) by several nanometers while preserving workpiece topography. HC-MRF accelerates surface finishing by integrating magnetorheological finishing (MRF) with chemical etching. A fluid that exhibits magnetorheology (MR), comprising abrasives, carrier fluid, stabilizing agents, and magnets, exhibits increased viscosity under magnetic field conditions, enabling the functionality of MRF. Furthermore, the improved magnetic resonance fluid and the relative motion between the pieces of metal produce the necessary level of surface roughness [8]. Additionally, there have been multiple attempts to improve the rheological properties of MR fluid to augment the efficiency of the magnetic field-assisted abrasive finishing (MFAF) process. In the weak magnetorheological shear thickening polishing fluid, Ming et al. [9] incorporated particles of polyhydroxy polymer into the thickening phase to improve zirconia ceramic surface quality. Finishing was faster and more effective than with standard MR fluid. Zhang et al. [10] made magnetic abrasive particles and iron-based silicon carbide composites. Upon completion, the surface roughness (R_a) of molded steel drops from 372 nm to 51 nm. The tool for polishing was created by Zhang et al. [11] in a separate study. They made it by layering silicon carbide on a permanent magnet. The sophisticated polishing tool, operating at a velocity of 15 km/h, reduced the roughness of the SS316 workpiece to 258 nm. Fan et al. [12] combined magnetorheological (MR) and shear thickening fluids and achieved 54 nm surface finish on titanium grade 2. Another important study reported the effectiveness of the same process strategy with a reduction in surface roughness from 465 nm to 35 nm on alumina tubes [13]. The optimized MR fluid composition effectively balances shear yield stress and abrasive action, resulting in uniform material removal without surface damage during MRF. Experimental results confirm a significant reduction in surface roughness while preserving surface integrity. The findings establish MRF as a highly efficient finishing technique for high-precision aluminum optical components [14].

This chapter focuses on describing how finishing procedures are chosen according to the material, desired surface quality, and application. Various polishing mechanisms were examined from mechanical, chemical, magnetic, optical, and ultrasonic perspectives (Fig. 4.1). New polishing processes were developed to provide smooth surfaces on aluminum alloys based on these phenomena. Figure 4.1 shows the general approaches, related to polishing, utilized to achieve smooth and defect-free surfaces on materials.

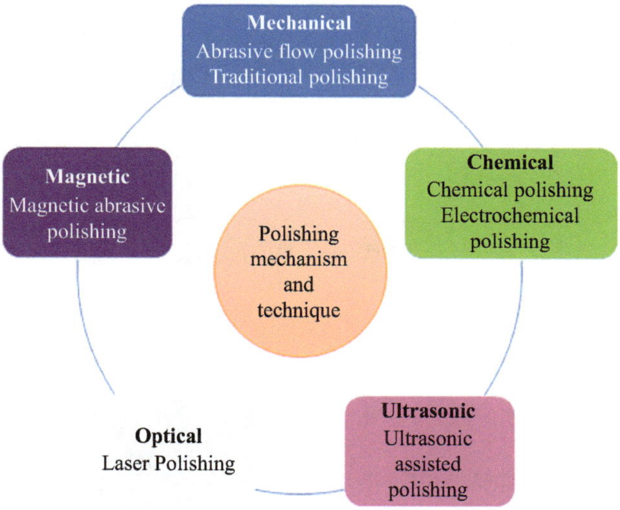

Fig. 4.1: Various polishing mechanisms and techniques to achieve finished surfaces.

4.2 Impact of Surface Roughness on Optical Application

The MRF using permanent-magnet-based tools is effective for finishing blind hole surfaces with controlled material removal. The magnetic field strength and tool geometry significantly influence finishing forces and surface roughness reduction. Experimental results show substantial improvement in internal surface finish without inducing subsurface damage [15]. The effects are contingent upon the magnitude and characteristics of the roughness in relation to the wavelength of light. Figure 4.2 illustrates characteristics of elements influencing the efficacy and quality of an optical component. Each adjacent box denotes a characteristic or flaw that affects the interaction of light with the optical material, hence influencing its precision, efficacy, and longevity.

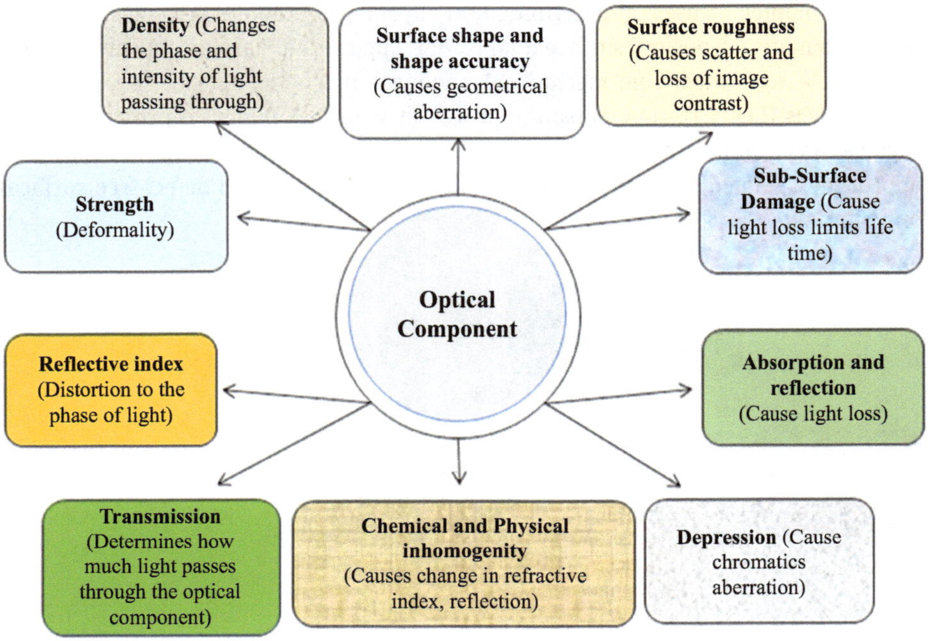

Fig. 4.2: Characteristics of optical components.

Some of the important aspects are discussed further.

4.2.1 Light Scattering and Diffusion

Smooth surfaces: When the roughness is significantly less than the wavelength of light (sub-wavelength roughness), the surface functions as an ideal mirror with negligible scattering.

Rough surfaces: When the roughness is equal to or exceeds the wavelength, considerable scattering transpires, resulting in diffused or scattered reflections rather than specular reflections.

4.2.2 Reflection and Transmission

Elevated roughness: Enhanced roughness results in augmented diffuse reflection and diminished specular reflection, impacting applications such as mirrors and lenses.

Minimal roughness: Critical for high-precision optics, particularly in laser systems, where specular reflection is vital.

Effect on transparent substances: Surface roughness in optical windows or lenses can lead to light diffusion, hence diminishing transmission efficiency.

4.2.3 Deficits and Assimilation

Rough surfaces increase light absorption due to many scattering events, which is detrimental to fiber optics and laser mirrors. In semiconductor photonics, waveguide roughness may elevate optical losses.

4.2.4 Diffraction and Interference

Surface roughness in precision optical components, such as gratings and interferometers, can induce diffraction phenomena and optical distortions.

4.2.5 Influenced Applications

Laser optics necessitate exceptionally smooth surfaces to minimize dispersion and facilitate the transmission of coherent light. To preserve image quality, the mirrors and lenses of microscopes and telescopes must be meticulously polished.

Display technology: Anti-glare screens and optical materials employ regulated roughness. The roughness of waveguides affects the efficiency of signal transmission in photonic devices.

4.3 Surface Finishing Techniques

The prevalent utilization of modern engineering materials, recognized for their remarkable hardness, strength, and toughness, has presented novel obstacles in machining operations within contemporary industrial contexts. Conventional procedures frequently encounter difficulties with such materials, necessitating alternative approaches that are not impeded by the mechanical qualities of the workpiece. The quality of surface finish is a critical determinant, since it directly influences component performance, encompassing energy efficiency, wear resistance, and corrosion protection. While finishing operations are crucial for achieving quality standards, they are often labor-intensive and substantially increase manufacturing costs. The ongoing advancement of manufacturing technology has led to an increasing demand for the processing of intricate and high-performance materials, including ceramics, austenitic steels, stainless steels (SUS), and various non-ferrous alloys such as aluminum, brass, bronze, copper, and gunmetal. These materials present challenges for traditional finishing techniques such as grinding, lapping, and honing. As a result, there is a significant movement toward the adoption of unconventional or nontraditional or advanced finishing techniques. These advanced processes can produce high-precision surfaces and stringent dimensional tolerances, particularly in industries belonging to aerospace, biomedical, and micromanufacturing sectors, requiring complex geometries, enhanced material compatibility, and superior surface properties [16]. Methods of finishing can be characterized in two modules as traditional and advanced finishing techniques as shown in Fig. 4.3.

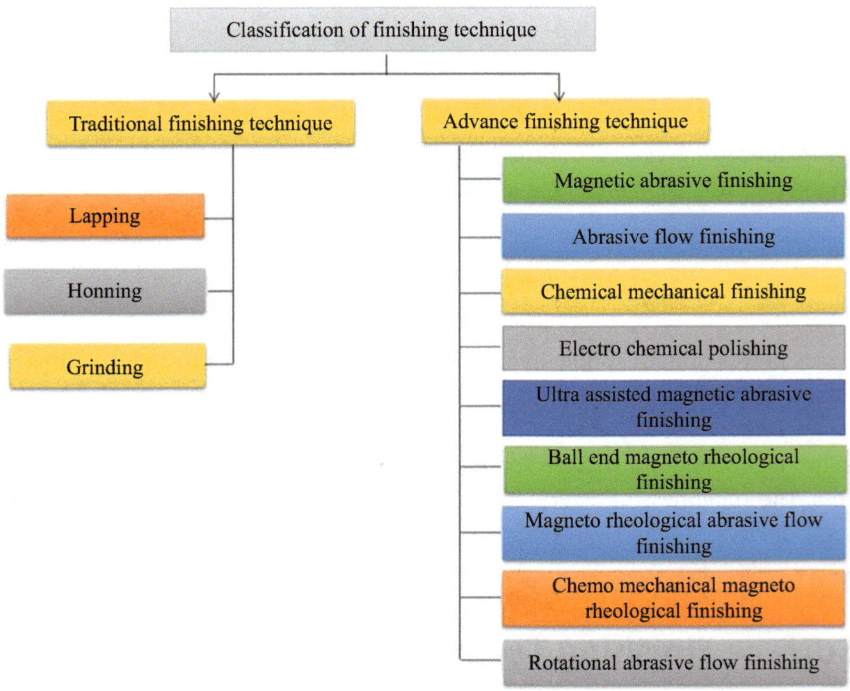

Fig. 4.3: Classification of finishing techniques.

Traditional techniques such as lapping, honing, and grinding are commonly employed for surface enhancement, although they possess constraints in attaining ultra-precision finishes. Conversely, advanced techniques utilize contemporary processes including magnetic abrasive finishing, abrasive flow finishing, chemical mechanical finishing, electrochemical polishing, and various hybrid methods such as ultra-assisted magnetic abrasive finishing, ball-end MRF, magnetorheological abrasive flow finishing, chemo-mechanical MRF, and rotational abrasive flow finishing. These sophisticated methods provide enhanced surface quality, improved control over material removal, and application to intricate geometries.

Chemical mechanical polishing (CMP) is extensively employed worldwide to achieve atomically clean surfaces. Conventional CMP techniques frequently employ hazardous and caustic slurries, posing environmental risks. Green CMP technology has been developed to mitigate this impact. These techniques are intended for materials such as copper, alloys, sapphire, fused silica, silicon, and diamond. Green CMP attains superior surface finishes while markedly reducing environmental hazards. These developments are particularly beneficial in sectors such as semiconductors, optoelectronics, and microelectronics. In summary, green CMP provides an eco-friendly substitute for traditional polishing methods [16].

4.3.1 Chemical Mechanical Polishing (CMP)

Aluminum-based alloys are extensively utilized in industries such as defense, aerospace, electronics, chemicals, and equipment due to their low electrical resistivity, superior thermal conductivity, strength, manufacturability, and compatibility with silica and silicon. In applications requiring exceptional surface quality such as aluminum gates and circuit boards, polishing is essential for minimizing surface roughness. CMP is acknowledged as an exceptionally efficient approach for attaining smooth and planar metallic surfaces. CMP attains a refined workpiece surface by the integration of mechanical abrasion and chemical etching [17, 18]. The continuous shrinking of integrated circuit characteristics has propelled the advancement of CMP technology, creating new prospects for sophisticated semiconductor fabrication. Figure 4.4 presents the schematic representation of CMP.

Having determined that the CMP of aluminum encompasses both passivation and direct material removal methods [19], an optimized balance between chemical reactions (oxidation and passivation) and mechanical abrasion is essential for effective material removal. The results demonstrate that alumina-based slurry composition strongly governs surface roughness reduction and polishing efficiency. However, the work also notes the inherent difficulty in achieving true mirror-like finishes on aluminum due to its reactive and soft surface nature [20]. The efficacy of the CMP process primarily relies on sustaining a harmonious interplay between chemical etching and mechanical polishing. Standard CMP slurry comprises abrasive particles, oxidizing agents, pH modifiers, corrosion inhibitors, and dispersants, with the composition significantly influencing overall efficacy [21]. Silica and alumina are common abrasive components in traditional CMP slurries. More recently, better slurries with hybrid abrasives, such as those with doped and core-shell structures, have been produced to make them even more effective. However, there is a lack of research on the application of hybrid abrasives that mix inorganic particles with two-dimensional layered nanomaterials, especially for polishing soft metals. These slurry formulations also frequently use classic acids such as nitric, sulfuric, and phosphoric acids, hence exacerbating environmental contamination. The advancement of environmentally sustainable and effective CMP slurries has been a persistent area of research.

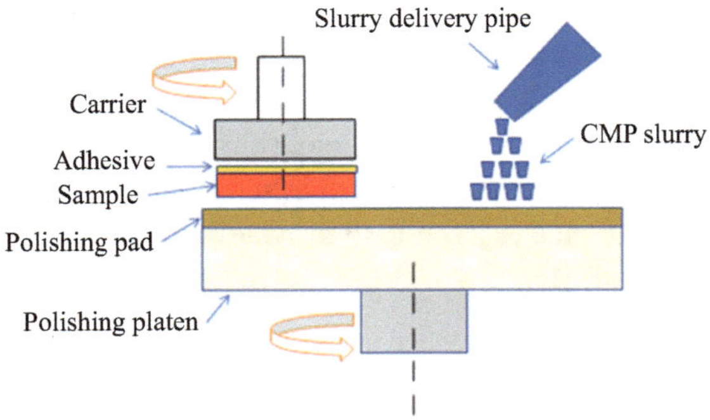

Fig. 4.4: Working principle of chemical mechanical polishing (CMP).

4.3.2 Magnetorheological Finishing (MRF)

Figure 4.5 shows the working principle of MRF process. MR fluids are intelligent suspensions made of magnetic particles scattered in a liquid medium on the scale of microns. These fluids change phase from liquid to semisolid when exposed to a magnetic field, producing a yield stress that is driven by the field. Polishing hard or brittle surfaces, whether flat or curved, is a common use of MRF. The use of magnetic resonance fluids allows for the efficient handling of different abrasive particles and the exact control of polishing forces by creating a viscoelastic contact between the workpiece and the magnetic field source [22]. An important study developed an optimized magnetic field-assisted finishing tool using magnetorheological fluid, designed through finite element simulations to enhance magnetic flux concentration in the polishing zone. The use of mu-metal for magnetic shielding improved flux control, and experimental validation confirmed the tool's effectiveness in achieving nanometer-level finishing (R_a value 10 nm) on titanium surfaces [23].

On the other hand, MRF is not the best choice for huge flat surfaces due to its small contact area and low polishing force. Consequently, a number of studies have concentrated on finding ways to increase the MRF processes contact area and polishing force. A cluster-type MRF disk with a dynamic magnetic field was devised by Huang et al. [24] to increase the tool-to-workpiece contact zone. Thanks to this improvement, large regions may be planarly finished on brittle and hard semiconductors such as silicon, sapphire, and silicon carbide.

Traditional wheel-type, cylindrical ball-head, and flat disk polishers often only work in single-shear mode, which severely restricts the amount of force that can be

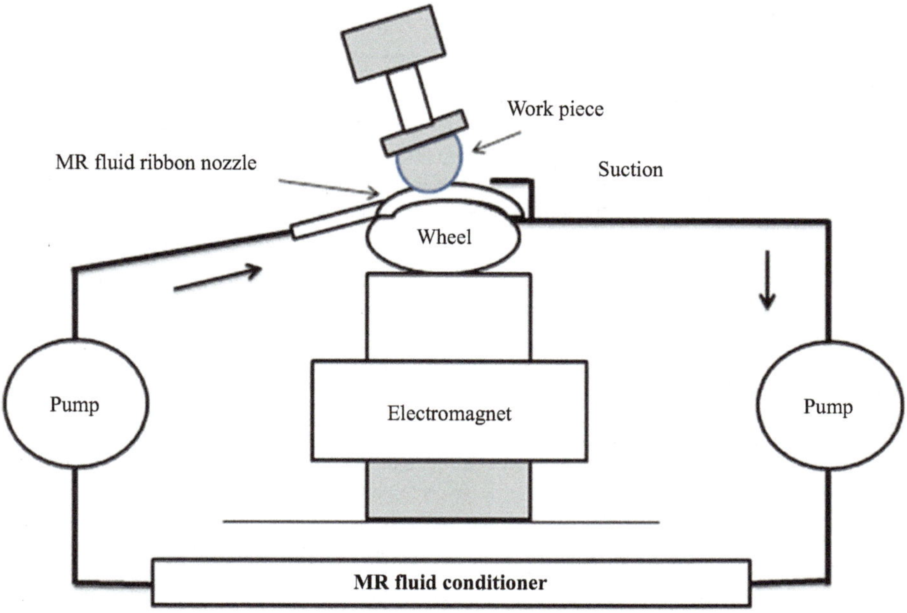

Fig. 4.5: Working principle of magnetorheological finishing (MRF).

applied and reduces their overall efficiency. MR fluids, on the other hand, show far larger compressive and tensile yield stresses than shear yield stress under the same magnetic field settings when used in compression-shear mode. According to Zhang et al. [25], shear yield stress was multiplied by ten after pre-compressing MR fluids. Wang et al. [26] also studied MR fluids' mechanical behavior in the same magnetic settings and came to the same conclusion: compressive yield stress was substantially higher than tensile stress, which was roughly four times higher than shear yield stress. Jain et al. [27] innovated a magnetorheological finishing approach combined with chemical action for polishing silicon surfaces. The process improves surface quality by controlling material removal through magnetic and chemical effects. It shows potential for achieving smooth and damage-free surfaces suitable for precision applications. Some important previous works on MRF of various materials are presented in Table 4.1.

Tab. 4.1: Literature survey on processing of different materials by MRF.

Author	Material	Finding
Deng et al. [28]	Al mirror	MRF was applied in conjunction with the small tool polishing technique to eliminate the diamond-turned surface's lingering figure errors. The rough surface was brought down to 3 nm and texture to RMS 0.025λ.

Tab. 4.1 (continued)

Author	Material	Finding
Yin et al. [29]	BK7 glass	In the beginning, an electrolytic in-process dressing grinding was carried out to bring the microroughness and figure accuracy down to a more manageable level. After performing MRF, it was shown that the form accuracy can be improved to a level of $\lambda/18$ P-V, and a microroughness of 0.56 nm RMS can be achieved in just 17 min.
Lormeau et al. [30]	Fused silica	MRF on advanced manufacturing systems has the capability to produce low roughness surfaces and high-precision freeform surfaces.
Catrin et al. [31]	Fused silica	MRF was found capable in removing surface and subsurface defects.
Beier et al. [32]	Electroless Ni-coated Al mirror	Polishing wheels with a diameter of 370 mm are used to effectively correct long-wave errors, while polishing wheels with a diameter of 50 mm are used to effectively correct fine figure errors. Both the shape accuracy and the microroughness were drastically improved.
Maloney et al. [33]	CaF_2 and Si	Low and middle spatial frequency errors can be corrected by combining small tools with C30 MR fluid. To achieve roughness <1.5 Å RMS suiting industrial and medical applications is possible.
Zhong et al. [34]	Electroless Ni	To fulfill the requirements of X-ray optical system application, the surface roughness was brought down to around 0.3 nm R_a. Cyclic residual marks or ripples were significantly reduced after being processed through MRF.
Piche et al. [35]	Fused silica	To efficiently correct the figure errors on optical surface, a new high-numerical-aperture MRF mode was utilized. Accuracy of shape on fused silica sphere was achieved at a level of about 5 nm RMS.
Fan et al. [36]	Ti alloy	Chemical-enhanced magnetorheological shear thickening polishing with eco-friendly slurry could achieve 56 nm surface roughness.
Zhong et al. [37]	CaF_2	It was possible to achieve RMS $\lambda/150$ and roughness Ra 0.3 nm that were significantly better than what produced during the basic manufacturing.

4.3.3 Chemo-mechanical Magnetorheological Finishing (CMMRF)

The advantages of both CMP and MRF are brought together in a hybrid technique called chemical-mechanical magnetorheological finishing, or CMMRF. It uses mechanical action in conjunction with chemical softening to remove material. By interacting chemically, the workpiece surface becomes more porous, which the MR polishing medium uses to adhere to and effectively remove the softened layers from surface protrusions. Using improved magnetic abrasive particles, enhancing system performance, incorporating hybrid techniques, and refining operational parameters are all ways to enhance the CMMRF process [38].

The polishing mechanism of MR fluid-based surface finishing relies on magnetic abrasives, which function as nanoscale cutting tools to eliminate surface flaws. Because of their relative simplicity of preparation and application, loosely blended magnetic abrasives have traditionally seen extensive use. With the progress of finishing, the material removal rate decreases due to the expulsion of these mixed abrasives caused by the considerable centrifugal forces generated by the revolving magnet. Inadequately mixed abrasives also pose a risk of becoming embedded in the metal, which can result in surface flaws. Mixed magnetic abrasives are less efficient and result in a more time-consuming finishing process. One viable option to overcome this restriction is the use of sintered magnetic abrasives. Sintered magnetic abrasives, consisting of bonded magnetic and abrasive particles, are extensively utilized for attaining exceptional surface finishes owing to their robust cutting capacity and magnetic responsiveness [39]. These abrasives create chain-like formations in magnetic fields, enhancing surface quality via shearing and fatigue while minimizing tool wear.

The significant advantages of CMP, MRF, and CMMRF are presented in Fig. 4.6.

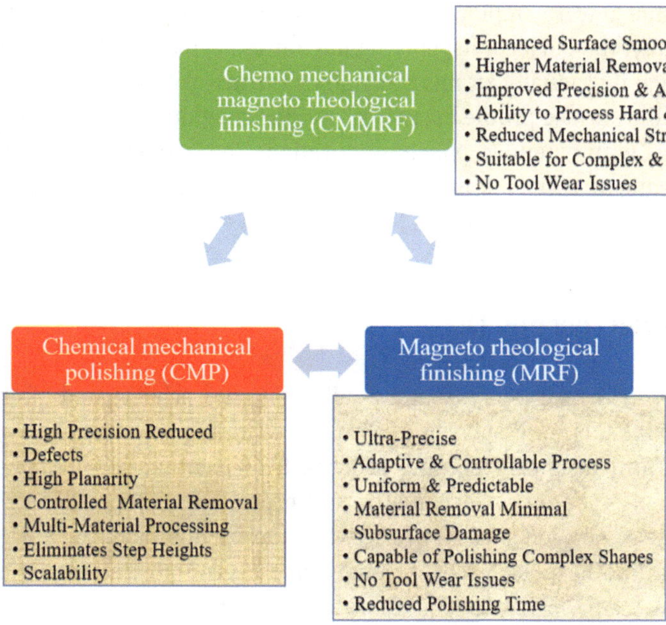

Fig. 4.6: Advantages of various hybrid finishing techniques.

4.4 Surface Characterization

To assess the efficacy and quality of a material's surface following finishing procedures such as HCMRF, surface finishing characterization is essential. Essential criteria encompass surface roughness (R_a), flatness, texture, subsurface damage, and material loss rate. These characteristics guarantee that the surface fulfills both functional and aesthetic criteria, especially in high-precision domains such as optics, electronics, biomedical implants, and aircraft components. Surface assessment is generally conducted utilizing tools such as 3D optical profilometers, atomic force microscopy (AFM), scanning electron microscopy (SEM), and interferometers. In addition to improving product performance and ensuring quality, precise characterization optimizes finishing process management [40].

4.4.1 3D Optical Profilometry

Figure 4.7 depicts the three-dimensional surface morphology, surface profile, and surface roughness of the aluminum alloy for different finishing durations [41]. Two measuring lines (line 1 and line 2) were employed to assess the surface profile and roughness, respectively. Initially, the surface displays significant imperfections like those previously noted, although it gradually attains a smoother texture as finishing advances. The deformation of the surface profile decreases, and the roughness curve exhibits diminished variations with extended polishing duration. The surface roughness markedly diminishes, reducing from 480 nm to 248 nm after 25 min of intensive finishing, and subsequently to 155 nm after 50 min [41].

4.4.2 Atomic Force Microscopy (AFM)

Figure 4.8 illustrates the AFM images of aluminum alloy samples prior to and after processing [42]. The initial surface displayed profound machining scratches and a hardened layer created during the machining process. SEM examination demonstrates a substantial enhancement in surface quality post-processing. Composite magnetic abrasives efficiently eradicated surface peaks, while the friction from the abrasive particles reduced grinding marks. The presence of fine composite magnetic abrasives in the finishing media facilitates the accurate elimination of surface imperfections. Prior research has indicated that employing finer abrasives in traditional mixed abrasive powders can effectively eliminate grinding marks during precision finishing [40–42].

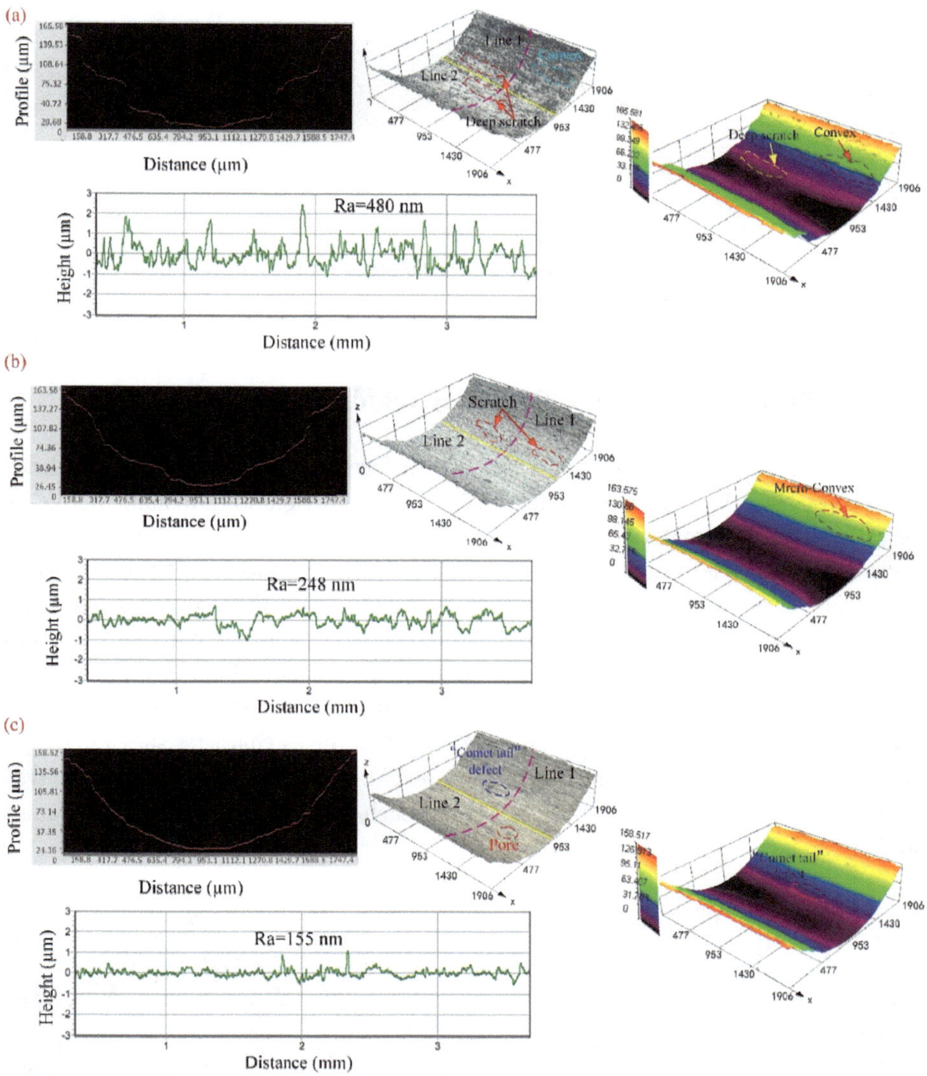

Fig. 4.7: 3D surface topography image of aluminum surface (a) before, (b) 25 min of finishing, and (c) after MRF process [41] (with kind permission from Elsevier).

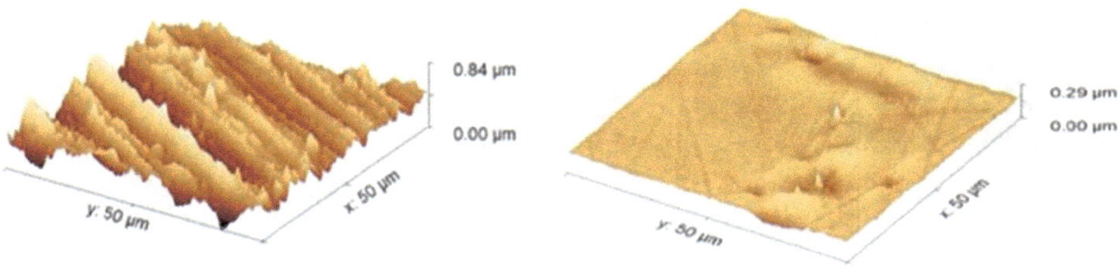

Fig. 4.8: AFM image of aluminum surface (a) before CMMRF process and (b) after CMMRF process [42] (with kind permission from Elsevier).

4.4.3 Scanning Electron Microscopy (SEM)

Figure 4.9 displays SEM images illustrating the surface state at various polishing phases [41]. Figure 4.9(a) depicts the unrefined surface exhibiting significant molding marks, severe abrasions, and material accumulation. Following 25 min of polishing, the protrusions on the workpiece surface have been diminished to microconvex structures, yielding a relatively smoother finish; yet, uneven scratches persist, as illustrated in Fig. 4.9(b). Figure 4.9(c) depicts the final polished condition, wherein abrasive particles constantly traverse the surface during processing, resulting in the creation of uniformly distributed fine scratches on the inner surface due to repeated contacts between particles and the surface. This procedure efficiently eliminates residual convex characteristics and irregular scratches, resulting in a smooth surface devoid of micropeaks and deep grooves. These findings validate that the MP-MSTP approach effectively generates high-quality surfaces on narrow aluminum alloy.

Fig. 4.9: SEM images of aluminum surface (a) before MRF, (b) after MRF, and (c) final polished surface [41] (with kind permission from Elsevier).

4.5 Conclusion

This chapter demonstrated the effectiveness of HCMRF as a promising solution to address the limitations of conventional finishing methods such as grinding and lapping for the required finishing of materials for optical applications. By integrating the deterministic mechanical action of MRF with the chemical softening mechanism of CMP, HCMRF enables controlled, efficient, and uniform material removal. The MR fluid, whose rheological properties are governed by an external magnetic field, provides precise control over finishing forces, making the process suitable for intricate geometries and freeform optical surfaces. Simultaneously, chemical interaction facilitates localized surface softening, enhancing finishing efficiency while minimizing subsurface damage.

The review of existing literature confirmed that MR-based processes can achieve significant reductions in surface roughness and improvements in form accuracy for optical materials, including aluminum alloys and other advanced substrates. Discussion on surface characterization techniques such as 3D optical profilometry, AFM, and SEM validated the capability of hybrid processes to eliminate surface irregularities and produce uniform, defect-free finishes suitable for optical applications.

Furthermore, it was realized that process parameters including work gap, slurry flow rate, feed rate, abrasive concentration, and finishing time play a crucial role in determining surface quality and material removal rate. Optimizing these parameters can ensure enhanced finishing efficiency and prevent surface deterioration. The use of composite and sintered magnetic abrasives further improved cutting performance and magnetic responsiveness.

Overall, HCMRF has emerged as a versatile and controllable finishing technique capable of meeting the stringent requirements of modern optical manufacturing. With further advancements in eco-friendly slurries, magnetic tool design, and process automation, HCMRF holds strong potential for next-generation ultra-precision surface finishing applications.

List of Acronyms and Nomenclature

CIP	Carbonyl iron particle
MRF	Magnetorheological finishing
MFAF	Magnetic field-assisted abrasive finishing
HCMRF	Hybrid chemo-magnetorheological finishing
CMP	Chemical mechanical polishing
AFM	Abrasive flow finishing
SMA	Simply mixed abrasive
EIP	Electrolyte polishing
AFM	Atomic force microscopy

RMS	Root mean square
SEM	Scanning electron microscopy
MHS	Microwave hybrid sintering
R_a	Average surface roughness
SiC	Silicon carbide
RPM	Revolution per minute
MRR	Material removal rate
Al	Aluminum

References

[1] Wen Y, Hu J, Pagilla PR. A novel robotic system for finishing of freeform surfaces. In Proceedings of the IEEE International Conference on Robotics and Automation. 2019 May. p. 5571–5577.

[2] Pan R, Zhao W, Wang Z, Ji S, Gao X, Chen D, et al. Research on an evaluation model for the working stiffness of a robot-assisted bonnet polishing system. J Manuf Process. 2021;65:134–143.

[3] Sah MK, Vijaya A. Optimizing highly reflective CNC abrasive polishing on aluminum sheets. Mater Manuf Process. 2024;39(8):1144–1161.

[4] Saif M, Rawat RK. Investigation of aluminum alloy 6061 in wire-EDM regarding surface roughness and material removal rate by adopting optimization techniques. J Mech Eng Sci. 2023;9410–9420.

[5] Huang Y, Wang C, Ding F, Yang Y, Zhang T, He X, et al. Principle, process, and application of metal plasma electrolytic polishing: A review. Int J Adv Manuf Technol. 2021;114:1893–1912.

[6] Sakthivelu S, Anandaraj T, Selwin M. Multi-objective optimization of machining conditions on surface roughness and MRR during CNC end milling of aluminium alloy 7075 using Taguchi design of experiments. Mech Mech Eng. 2017;21(1):95–103.

[7] Nesli S, Yilmaz O. Surface characteristics of laser polished Ti-6Al-4V parts produced by electron beam melting additive manufacturing process. Int J Adv Manuf Technol. 2021;114:271–289.

[8] Rajput AS, Singh A, Kapil S, Das M. Investigations on the trochoidal toolpath for processing the biomaterial through magnetorheological fluid assisted finishing process. J Manuf Process. 2022;76:812–827.

[9] Ming Y, Huang X, Zhou D, Li X. A novel Non-Newtonian fluid polishing technique for zirconia ceramics based on the weak magnetorheological strengthening thickening effect. Ceram Int. 2022;48(5):7192–7203.

[10] Zhang G, Zhao Y, Zhao D, Zuo D, Yin F. New iron-based SiC spherical composite magnetic abrasive for magnetic abrasive finishing. Chin J Mech Eng. 2013;26:377–383.

[11] Zhang J, Wang H, Senthil Kumar A, Jin M. Experimental and theoretical study of internal finishing by a novel magnetically driven polishing tool. Int J Mach Tools Manuf. 2020;153:103552.

[12] Fan Z, Tian Y, Zhou Q, Shi C. Enhanced magnetic abrasive finishing of Ti–6Al–4V using shear thickening fluids additives. Precis Eng. 2020;64:300–306.

[13] Yang Z, Fan Z, Zhang X, et al. Experimental investigations on chemical-assisted magnetorheological shear thickening polishing for alumina tubes. Int J Adv Manuf Technol. 2025;141:5315–5327. https://doi.org/10.1007/s00017-025-16973-y.

[14] Jie G, Chao Y, Changxi X, Peng S. Material removal mechanism and MR fluid for magnetorheological finishing of an RSA-6061 aluminum alloy mirror. Appl Opt. 2022;61:10098–10104.

[15] Sirwal SA, Singh AK, Paswan SK. Experimental analysis of magnetorheological finishing of blind hole surfaces using permanent magnet designed tools. J Braz Soc Mech Sci Eng. 2020;42:1–23.

[16] Sah MK, Singh H. A review of surface finishing by magnetic abrasive finishing: Challenges and prospects. J Micro Manuf. 2026;1–33. doi: 10.1177/25165984251379711.

[17] Zhang ZY, Liao LX, Wang XZ, Xie WX, Guo DM. Development of a novel chemical mechanical polishing slurry and its polishing mechanisms on a nickel alloy. Appl Surf Sci. 2020;506:9.
[18] Wang H, Cheng B. Study of a magnetorheological brake under compression-shear mode. Smart Mater Struct. 2020;29(1):017001.
[19] Zhang Z, Zhang W, Zhang S, Li F. Chemical mechanical polishing of aluminum alloys using alumina-based slurry. Tribology. 2019;39(1):109–117.
[20] Li T, Sun H, Wang D, Huang J, Li D, Lei F, Sun D. High-performance chemical mechanical polishing slurry for aluminum alloy using hybrid abrasives of zirconium phosphate and alumina. Appl Surf Sci. 2021;537:147859.
[21] Wang YL, Tseng WT, Chang SC. Chemical–mechanical polish of aluminum alloy thin films: slurry chemistries and polish mechanisms. Thin Solid Films. 2005;474(1–2):36–43.
[22] Kumar M, Yadav HN, Kumar A, et al. An overview of magnetorheological polishing fluid applied in nano-finishing of components. J Micro Manuf. 2022;5(2):82–100.
[23] Barman A, Das M. Design and fabrication of a novel polishing tool for finishing freeform surfaces in magnetic field assisted finishing (MFAF) process. Precis Eng. 2017;49:61–68.
[24] Huang Z, Yan Q, Pan, Chen Z, Lu J. Dynamic magnetic field magnetorheological finishing with constant load and variable gap. *Precis Eng*. 2024;*86*:388–399.
[25] Xz Z, Xl G, Pq Z. Study on the mechanism of the squeeze-strengthen effect in magnetorheological fluids. J Appl Phys. 2004;96(4):2359–2364.
[26] Wang H, Bi C, Kan J. The mechanical property of magnetorheological fluid under compression, elongation, and shearing. J Intell Mater Syst Struct. 2011;22(8):811–816.
[27] Jain VK, Ranjan P, Suri VK, Komanduri R. Chemo-mechanical magneto-rheological finishing (CMMRF) of silicon for microelectronics applications. CIRP Ann Manuf Technol. 2010;59(1):323–328.
[28] Deng J, Peng X, Hu H, Ge K. Study on combined polishing process of aspherical aluminum mirrors. Proc SPIE. 2017;10460:104600A.
[29] Yin S, Ohmori H, Uehara Y, et al. Ultra-precision finishing process integrated ELID- grinding and MRF for BK glass. Proc SPIE. 2009;7282:72821.
[30] Lormeau JP, Supranowitz C, Dumas P, et al. Field-proven technologies for fabrication of high-precision aspheric and freeform optical surfaces. Proc SPIE. 2015;9442:944203.
[31] Catrin R, Neauport J, Taroux D, et al. Magnetorheological finishing for removing surface and subsurface defects of fused silica optics. Opt Eng. 2014;53:092010.
[32] Beier M, Hartung J, Kinast J, et al. Fabrication of metal mirror modules for snap-together VIS telescopes. Proc SPIE. 2015;9633:963313.
[33] Maloney C, Lormeau JP, Dumas P. Improving low, mid and high-spatial frequency errors on advanced aspherical and freeform optics with MRF. Proc SPIE. 2016;10009:100090.
[34] Zhong X, Hou X, Yang J. Super-smooth processing X-ray telescope application research based on the magnetorheological finishing (MRF) technology. Proc SPIE. 2016;9963:99630.
[35] Piche F, Maloney C, VanKerkhove S, et al. New high-precision deep concave optical surface manufacturing capability. Proc SPIE. 2017;10448. doi: 10.1117/12.2279809.
[36] Zenghua F, Zihao Y, Xiang Z, Jun G, Yebing T. Material removal behavior and surface evolution in chemical-enhanced magnetorheological shear thickening polishing of titanium alloy slender tubes. Tribol Int. 2026;218:111667.
[37] Zhong X, Fan B, Wu F. High precision processing of CaF_2 application research based on the magnetorheological finishing (MRF) technology. Proc SPIE. 2017;10448:104481Y.
[38] Kumar Y, Singh H. Chemomechanical magnetorheological finishing: process mechanism, research trends, challenges, and opportunities in surface finishing. J Micro Manuf. 2021;5(2):193–206.
[39] Chen Y, Zhang MM, Liu ZQ. Study on sintering process of magnetic abrasive particles. Adv Mater Res. 2011;337:163–167.

[40] Ghosh G, Sidpara A, Bandyopadhyay P. Review of several precision finishing processes for optics manufacturing. J Micro Manuf. 2018;1(2):170–188.

[41] Li J, Fan Z, Gao J, Yang Z, Tian Y. Multi-pole magnetorheological shear thickening polishing on inner surface of aluminum alloy slender tubes. J Mater Res Technol. 2023;27:8258–8270.

[42] Kumar Y, Singh H. Experimental investigations on chemo-mechanical magnetorheological finishing of Al-6061 alloy using composite magnetic abrasive (CIP– Al_2O_3) developed via microwave-sintering route. J Manuf Process. 2023;99:765–780.

Orhan Çakır, Çağla Özbilek, Beyzanur Yanık, Miraç Mutlu, Salih Yeşilsancak, Mohammad Zaher Akkad

5 Casting-Based Hybrid Manufacturing Processes

Abstract: Casting-based hybrid manufacturing integrates traditional casting techniques with advanced processes such as machining, additive manufacturing, welding, forging, and external field assistance to overcome the inherent limitations of standalone casting. This chapter provides a comprehensive overview of such hybrid combinations, aiming to enhance dimensional accuracy, surface quality, mechanical performance, and production efficiency. Sequential hybridization with machining addresses casting defects and enables high-precision features, while additive-assisted casting improves mold complexity and customization, especially through methods such as hybrid investment casting and 3D sand printing. Welding and friction stir welding offer reinforcement and post-casting modification capabilities. Moreover, forging-casting hybridization, such as the integrated casting and forging process, yield improved microstructural uniformity and strength. Emerging assisted techniques including ultrasonic, electromagnetic, magnetic field, and vacuum-assisted casting introduce novel ways to refine grain structure, reduce porosity, and tailor material properties. These technologies, when combined strategically, facilitate the production of high-performance components in critical sectors such as aerospace, automotive, and biomedical engineering. This chapter ends with the discussion on environmental sustainability and ethical considerations, emphasizing reduced material waste, energy efficiency, and the need for inclusive adaptation to advanced manufacturing. Overall, casting-based hybrid manufacturing represents a transformative strategy for achieving next-generation manufacturing goals under Industry 4.0.

Keywords: additive manufacturing, casting, grain structure, hybrid manufacturing, machining, porosity

5.1 Introduction

Casting is a manufacturing process in which metal is melted and poured into a mold to get a certain shape, offering the advantage of producing both simple and complex geometries. In the future, it is expected that castings will be required in smaller sizes and more complex shapes, with reduced wall thickness and higher dimensional precision [1]. Casting is widely used in product development and mass production, with approximately 90% of all products being manufactured through this process. How-

 | https://doi.org/10.1515/9783111704456-005

ever, in recent years, the growth of the casting industry has tended to stagnate [2]. However, the traditional casting process entails multiple intricate and time-consuming steps, often requiring expensive tooling. Producing molds, cores, or patterns for limited production volumes is typically not economically viable, making conventional casting less suitable for low-batch or customized manufacturing applications [1]. Globalization, the need for innovation, and the drive to improve production efficiency present significant challenges for the metal casting sector. In the era of Industry 4.0, continuous research and development (R&D) activities are required to keep the casting technology competitive by enabling the shortening of production times and the improvement of the process's efficiency.

Hybrid manufacturing aims to integrate two or more distinct manufacturing processes within a single system to leverage the individual strengths of each method while mitigating their inherent limitations [3]. This synergistic approach enables enhanced flexibility, improved efficiency, and superior product quality compared to conventional standalone processes. Hybrid manufacturing processes can be mainly categorized based on the nature of process integration on two categories [4]. In the first group, known as simultaneous or synergetic hybrid processes, two or more energy sources or tools act concurrently within the processing zone to produce a combined effect. This category can be further subdivided into assisted hybrid processes, where one energy input enhances the effect of another, and mixed or combined processes, where the contributions of each source are distinct yet concurrent. The second group includes sequential hybrid processes, where multiple conventional manufacturing methods are applied in a controlled sequence, each contributing a distinct processing effect that is typically achieved in isolation in traditional workflows. Casting-based hybrid manufacturing involves sequentially combining casting with processes like machining, forging, or additive manufacturing (AM); and assisted casting processes in which assistance of heat, vibration, magnetic-field, and vacuum etc., to enhance casting mechanisms and/or to eliminate deficiencies. Overall, hybridization helps to improve cast part quality, minimize defects, and enhance productivity. Studies have demonstrated notable enhancements in dimensional accuracy, surface finish, and functional performance, making this approach suitable for high-precision applications [1, 2].

Overall, casting remains indispensable, yet escalating demands for complex, precise, low-volume parts and Industry 4.0 imperatives are challenging its traditional workflow. Hybrid manufacturing, particularly sequential casting-based variants address these challenges by merging complementary processes, shortening lead times, curbing costs, and elevating dimensional and functional quality. Consequently, hybridized casting technologies are poised to sustain competitiveness and unlock new high-performance applications across diverse industrial sectors.

5.2 Casting-Based Hybrid Manufacturing Processes

5.2.1 Hybridization of Casting and Machining

The main reason for the hybrid combination of casting and machining methods is to optimize production processes by combining the advantages of both manufacturing techniques. Casting is a cost-effective method for quickly shaping complex geometries, but it has limitations such as porosity, microstructural defects, and lack of precision [5, 6]. On the other hand, machining methods are highly effective in terms of precision, surface quality, and the creation of specific geometric designs [3]. By combining these two methods, the advantages of the fast production of casting and the precision of machining can be utilized to produce stronger, more precise, and high-quality parts. This combination also reduces costs while improving production times and enables the production of complex geometries with less waste and greater efficiency. The combination of these methods offers an effective solution for high-performance and demand-driven production across a wide range of industries, from automotive to aerospace, medical devices, and energy systems.

The machinability of gray cast iron with varying microstructures, particularly focusing on the influence of pearlitic content on milling performance using carbide and ceramic tools, was investigated by Dias and Diniz [7]. They demonstrated that alloys with 100% pearlite exhibited significantly faster tool wear and higher cutting forces compared to alloys with 50% pearlite and 50% ferrite, emphasizing the critical role of microstructural control in cast iron machinability. The ceramic tools consistently outperformed carbide tools due to their superior wear resistance and reduced susceptibility to thermal cracking.

Eraslan et al. [8] explored the machinability characteristics of austempered ductile iron (ADI) and cast steel (CS) under eco-friendly milling conditions, including dry, conventional cutting fluid (CCF), and minimum quantity lubrication (MQL). Machining tests conducted with TiAlN coated carbide end mills revealed that ADI resulted in greater tool wear and higher cutting forces than CS due to its unique ausferritic microstructure and induced strain-hardening phenomena. Interestingly, dry machining conditions proved more favorable regarding cutting forces, surface roughness, and tool wear compared to CCF and MQL conditions, due to reduced thermal shock and stress fluctuations at tool edges during intermittent cutting. Microstructural analysis revealed a clear differentiation in subsurface microstructural integrity between ADI and CS under different lubrication conditions, highlighting how MQL conditions effectively reduced thermal-induced transformations and microstructural stability.

Finally, chemical milling of cast titanium alloys Ti-6Al-4V and Ti-6Al-2Sn-4Zr-2Mo in hydrofluoric-nitric acid solutions were studied by Sefer et al. [9] to characterize microstructural corrosion behavior influencing subsequent machining operations. Using atomic force microscopy and electrochemical methods, they identified accelerated corrosion of α-laths relative to β-layers due to microgalvanic interactions between microconstituents, which significantly affected surface integrity during subsequent machining steps. Faster corrosion rates and higher α/β height differences were measured, emphasizing the critical role of initial chemical milling in preparing surface conditions favorable for subsequent precision machining. Their results underline the necessity of controlling microstructural interactions during chemical milling to enhance machinability and maintain component integrity.

5.2.2 Hybridization of Casting and Welding

Combining casting and welding methods facilitates achieving economical near-net-shape cast parts with improved properties. Complex parts can be cast cost-effectively with minimal waste, then welded or additively deposited with additional material to boost strength and integrity [10]. This hybrid approach exploits casting's scalability and welding's precision, yielding components more reliably and efficiently than either process alone, often with superior mechanical properties [10, 11]. Hybrid cast-weld processes face challenges like mismatched microstructures, residual stress, and welding-induced defects. High carbon cast alloys, for example, can form brittle phases in heat-affected zones, requiring preheating or heat treatments to prevent cracking [12]. Optimization of welding parameters and careful heat control are critical to minimize porosity or lack-of-fusion defects and ensure a strong, defect-free bond in the hybrid component.

Senff et al. [13] examined a hybrid processing method used in the joining of aluminum casting alloys and steels. This method strengthens the joint area with friction stir welding (FSW) after casting, where a steel insert is added to the aluminum material. The combination of casting and FSW effectively combines the advantages of both methods, resulting in a strong bond between the steel insert and aluminum. Microstructural analyses revealed that no intermetallic phases formed and that steel particles dispersed in the aluminum matrix. Single-sided FSW successfully transferred most of the tensile force of the steel material, while the double-sided welding method allowed greater tensile force transfer but was more costly and limited in accessibility. The results showed that the hybrid method produced robust joints capable of withstanding high tensile forces and indicated the need for future hole pattern optimization and durability testing. This study highlights the effectiveness of the casting and FSW combination in aluminum and steel joining and emphasizes the potential benefits of hybrid processes.

In another study by Gouveia et al. [14] investigated welded joints produced on ductile cast iron (700 MPa UTS), targeting the optimization of joint performance through various heat treatment combinations. As a joining-assisted casting method, this study employed shielded metal arc welding with Inconel filler metal to bond cast iron specimens. The samples were subjected to preheating, post-weld heat treatment (PWHT), and nickel edge enrichment to explore different hybrid sequences. The best mechanical outcome was achieved in the group labeled "P0_P0_W_P700," where no preheating was used, but PWHT at 700 °C was applied. This group retained approximately 59.8% of the base metal's UTS (399.1 MPa) and 50.8% of its strain capacity, reflecting optimal transformation of the cast base to a jointed structure. Scanning electron microscopy (SEM) images of the welded zone revealed moderate diffusion of carbon nodules into the filler region and minimal ledeburite formation, indicating controlled metallurgical behavior through careful thermal management.

5.2.3 Hybridization of Additive Manufacturing and Casting for Tooling

The integration of casting and AM technologies is creating a significant transformation in the casting sector [1]. This approach, which combines traditional casting methods with AM technologies, offers advantages in terms of cost and time while enabling the casting of complex and customized geometries. In traditional investment casting (IC), mold and model/pattern creation is a critical step, as the mold and pattern must have high precision and appropriate geometric design to ensure successful casting. In this context, AM technologies reduce the challenges of traditional mold production by enabling the rapid and low-cost creation of complex shapes [15].

IC is a commonly preferred casting method to produce high-precision metal parts with complex geometries [2]. In this method, liquid metal is poured into a mold and shaped, typically using a wax pattern system. In traditional methods, these wax patterns are removed through an autoclaving process. Hybrid investment casting (HIC) is an innovative casting method that combines traditional casting methods with modern technologies. In this method, plastic models produced using fused deposition modeling (FDM) replace traditional wax patterns [2]. FDM enables the rapid and cost-effective creation of mold models using 3D printing technology. By integrating these plastic mold models into the casting process, liquid metal is poured into the mold cavity to achieve the desired shapes. The plastic molds are burned at high temperatures (approximately 850–1,050 °C), and the liquid metal is poured to create the final shapes. Since FDM-based plastic models are used instead of traditional wax patterns, the process can be completed more quickly. Furthermore, the HIC method offers greater control over the mechanical and metallurgical properties during the casting process. This results in high-precision, customizable casting parts that can be used in various industrial applications [16].

Numerous studies have been conducted on the HIC method in the literature [17–20]. These studies focus on how the use of plastic molds in casting processes provides advantages in performance compared to traditional methods. This approach has significant potential, especially in biomedical applications and industrial areas requiring precision casting. In the hybrid casting process, the burning of plastic molds at high temperatures during the casting of liquid metal ensures structural compatibility between the mold and metal.

Studies using this method show that the mechanical properties obtained from plastic molds can be comparable to or even superior to those obtained from traditional wax patterns. Additionally, this method aims to reduce casting times and costs. In various applications, the performance advantages of the HIC method over traditional methods and its effectiveness in terms of production times are emphasized. The methodology for developing the HIC process is illustrated in Fig. 5.1.

In an important study, Kumar et al. [2] presented a development framework for HIC based on the use of plastic molds developed by FDM process. The focus of the study was to highlight the applicability of the HIC method in biomedical applications and to demonstrate its production flexibility. According to experimental results, using layer thicknesses of 0.0070 mm or finer significantly improved the surface finish, and the use of "hollow" molds reduced the cracking issues of ceramic shells. It was observed that the selection of fine layers contributed to achieving smooth surfaces by minimizing the staircase effect. Moreover, the "hollow" molds provided softening at lower temperatures due to their ability to collapse inward before reaching critical stress levels, which reduced the risk of cracking. These findings show that the HIC method offers production flexibility in industrial and biomedical applications and provides significant advantages in the precise production of various components.

Another experimental study [20] presents a digitally integrated methodology for the fabrication of titanium dental crowns by coupling laser-based geometric acquisition, numerical casting simulation, and rapid prototyping within the IC workflow. Crown geometries were digitized via 3D laser scanning and converted into stereolithography file format CAD models, which were used both for finite difference-based casting simulations in MAGMASOFT and for fabricating wax patterns through an FDM system. Simulation of mold filling and solidification enabled the optimization of runner and gating configurations to minimize shrinkage porosity. Subsequently, silicone molds were produced from RP patterns to enable mass replication of wax crowns, which were cast using centrifugal IC. Radiographic inspection confirmed defect-free castings with high surface integrity and dimensional fidelity. The study underscores the viability of this hybrid digital–casting framework in producing high-quality titanium prostheses and highlights its potential to replace conventional impression-waxing techniques, particularly by enabling predictive control of porosity and enhancing geometric reproducibility.

Figure 5.2 presents the integration of AM techniques across various stages of the casting process. It begins with model creation and pattern generation using stereolithography (SLA), FDM, or SLS, followed by mold and core fabrication. Wax shapes and ceramic shells are then developed using AM-assisted methods. The metal casting is performed via SLM or DLMS, with post-casting enhancements achieved through AM-based surface or dimensional refinements.

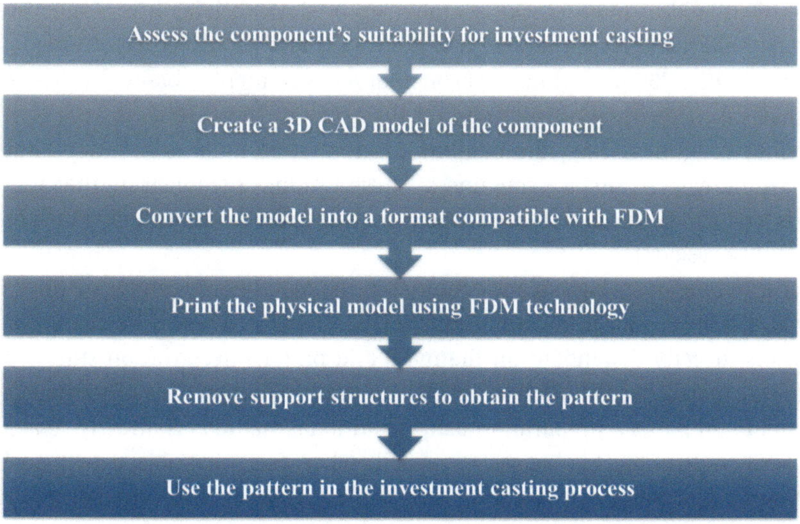

Fig. 5.1: Steps for developing HIC processes.

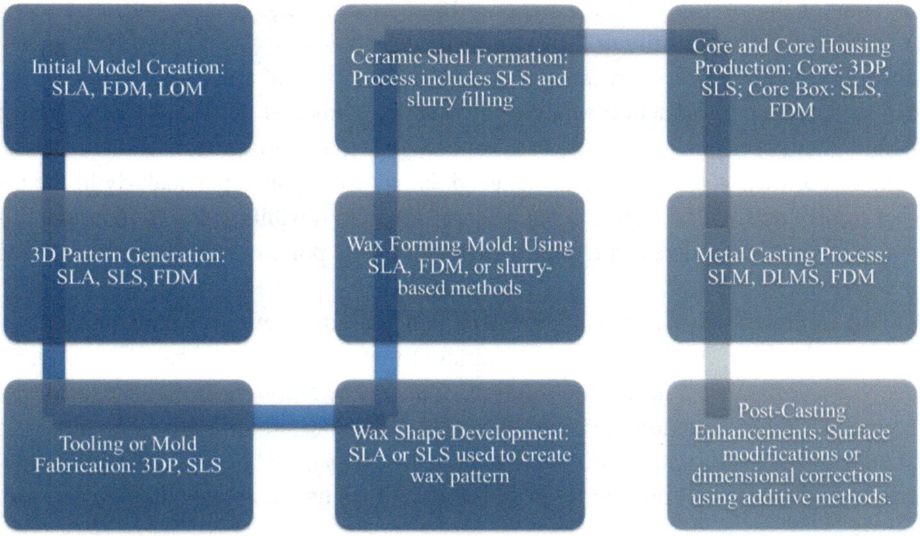

Fig. 5.2: Application of additive manufacturing in different phases of the casting process.

Kumar et al. [18] examined the FDM-assisted precision casting process in terms of biomedical implant production and optimized process parameters to reduce casting surface roughness. In the study, the first type of ABS plastic mold was coated with a thin layer of wax, while this coating was not used in the second type. The results showed that the wax coating improved the surface roughness of the casting by 67.63% and reduced the likelihood of mold cracking. Under optimal conditions, it was theoretically calculated that the surface roughness of the wax-coated molds, when oriented at a 0° angle, would be 1.899 μm, while experimental measurements obtained values of 1.912 and 1.890 μm. The study also found that the effect of the volume-to-area (*V*/*A*) ratio on surface roughness was minimal, but an increase in mold orientation angle resulted in an increase in surface roughness. These findings demonstrate that wax-coated molds perform better than those without coating and provide valuable insights for guiding future research.

Rapid prototyping holds significant potential in the metal casting industry for enhancing mold efficiency and production quality [16]. The effectiveness of metal casting processes is directly dependent on the mold and pattern preparation stages, as these processes typically account for about 70% of the production time. The time required for mold and pattern preparation using traditional methods is directly related to the complexity of the object being cast. Especially when designing objects with complex shapes and sizes, pattern preparation time can increase significantly, negatively affecting the production process in terms of time. In this context, increased preparation times, depending on the complexity of the patterns, reduce the efficiency of the casting process and increase the likelihood of casting defects. Therefore, pattern making, one of the first and critical steps in the casting process, is a stage that directly influences the final quality of the manufacturing. Integrating rapid prototyping technologies into these processes can overcome the limitations of traditional methods, significantly reduce production times while enable molds and patterns to be designed with higher precision. As a result, both time savings in the production chain and minimization of casting defects can be achieved. Rapid prototyping, particularly in the design of products with complex geometries, offers high flexibility, precise manufacturing, and cost advantages, carrying transformational potential in the metal casting industry [2].

3D sand printing (3DSP) is an AM technology used in foundries to produce sand molds and cores [21]. Compared to traditional mold production methods, it offers a faster, cost-effective, and flexible solution. 3DSP is also a preferred method for prototyping and the production of custom-designed parts. More than 50% of foundries use this technology to produce sand molds and cores. It especially provides a quick solution for new designs and custom production requirements, accelerating design processes.

However, there are also some challenges. High unit costs and low demand are two key factors that make the adoption of 3DSP technology in foundries difficult. Unlike traditional mold production methods, 3DSP eliminates mold costs, enabling foundries to achieve more profitable results in low-volume production. Despite this, limited demand and access issues prevent the widespread adoption of this technology. As a result, while 3DSP is an effective method that accelerates foundry processes and reduces costs, strategic planning is required to overcome the challenges [21].

The increasing complexity of metal parts for industrial and critical mission applications necessitates the adoption of new technologies. 3DSP, one of the layered manufacturing methods, addresses this issue by directly printing sand molds layer by layer. Using binder jetting technology, 3DSP offers solutions based on CAD files and selective binders, eliminating the need for mold tools. Figure 5.3 illustrates the research framework advancing casting applications by taking advantage of the freedom of design in 3DSP.

Traditional sand-casting methods face issues such as long preparation times, high costs, and limited design flexibility. In this context, 3DSP technology produces complex mold designs using direct printing methods and reduces defects. Defects caused by the oxidation of liquid metal can affect casting performance [23]. Therefore, the design of the sprue, which controls the flow rate of the liquid metal, is a critical factor. In one study, complex designs like the conical-helix sprue (CHSC) were optimized using 3DSP [22]. As a result of the experiments, the CHSC design reduced defects by 99.5% compared to the traditional straight sprue design and increased tensile strength by 8.4%. These findings demonstrate that casting quality can be significantly improved through 3DSP technology and new design methods. When combined with hybrid mold methods, these technologies offer innovative solutions in industrial applications.

In their studies, Krochmal et al. [24] compared the properties of two different materials produced from AlSi10Mg alloy and evaluated the mechanical properties and microstructures of hybrid joints by combining materials produced by traditional casting and laser-assisted powder bed fusion (AM) methods. The hybrid manufacturing method significantly improved the mechanical properties of the components obtained by combining the casting and AM processes. In hybrid samples, the weak effects of the casting region were strengthened by the integration of the AM material's microstructure, resulting in finer grain structures and higher hardness. For example, components obtained through hybrid manufacturing showed better low cycle fatigue (LCF) resistance up to 120 MPa. Compared to casting materials, smoother fracture surfaces were observed in hybrid joints, and crack growth was slowed. The study emphasizes that hybrid manufacturing is an effective method for enhancing material performance by strengthening weak areas in the casting section. Additionally, it was suggested that post-processing like heat treatments could further improve hybrid manufacturing processes and strengthen mechanical properties. These findings demonstrate that casting-based hybrid manufacturing methods offer a cost-effective and efficient solution for producing large components and improving material performance.

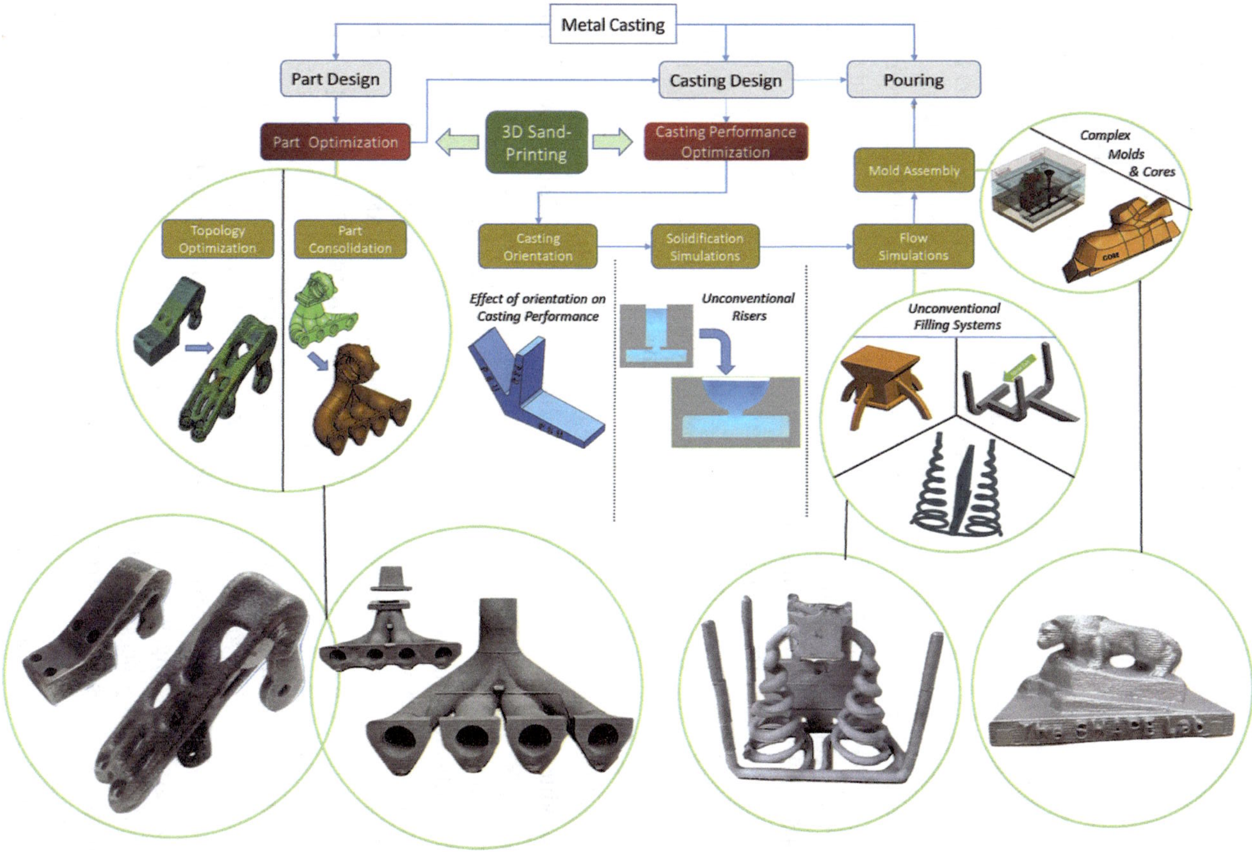

Fig. 5.3: Framework for manufacturing parts using 3D sand printing [22] (with kind permission from Elsevier).

5.2.4 Hybrid Casting and Forging

Casting-forging hybrid shaping technology is a method that combines the advantages of casting and forging processes. This technology is ideal for producing parts with complex shapes, as casting allows for easy production of complex shapes. By replacing extruded rods with cast rods as forging stock, a new, low-cost, and more efficient production route can be established, offering potential savings of 20% on a material level [25]. However, the availability of cast forging stock in preferred dimensions is often limited, primarily due to the high capital investment required for horizontal casting lines, which typically yield low production volumes. Additionally, cast materials commonly develop a surface segregation zone with a chemical composition that differs from the core. This zone must be removed through scalping before forging, leading to increased material waste and additional processing costs [25].

As a result, casting-forging hybrid shaping technology offers advantages such as reducing production costs, saving energy, and producing high-performance parts, while also presenting technical challenges and the need for appropriate material selection [26].

The integrated casting and forging process (ICFP) is a new manufacturing method that combines the advantages of casting and forging techniques [27]. Figure 5.4 illustrates the working principle of ICFP process. This process integrates both methods to produce complex shapes and achieve high mechanical properties. Initially, casting is performed (Fig. 5.4a, b), and then the final product is shaped using the forging process (Fig. 5.4c, d). ICFP enables the benefits of both methods to be utilized by combining the casting and forging processes.

A study on the ICFP aimed to make the production of complex-shaped and high-quality parts more efficient by combining the advantages of casting and forging processes [27]. Simulation and experimental results showed that an increase in casting temperature resulted in a longer waiting time for forging and a decrease in shaping force with higher casting temperatures. Moreover, as the static pressure of the injection piston and forging speed increased, the maximum shaping force and pressure of the parts increased. With the appropriate parameters identified in the study, ICFP allowed the production of parts with more complex shapes and high mechanical properties. The advantages of ICFP include fewer processing steps, energy efficiency, and cost-effectiveness. Additionally, the microstructure obtained in the forging region was finer and more uniform compared to the casting region, resulting in significant improvements in mechanical properties. Consequently, the ICFP method offers an effective solution for producing high-quality, complex-shaped parts and has the potential to replace traditional methods in industrial production. The working principle of ICFP is to combine casting and forging processes to efficiently use the latent heat from casting in the forging stage. To overcome the challenges encountered in traditional casting and forging methods, optimized configurations of ICFP were determined through trial-and-error methods, although this process can be inefficient. Due to the difficulty in measuring temperature and pressure distributions, the finite element method (FEM) was used to simulate these parameters in the study.

Chen et al. [28] investigated the grain refinement of as-cast AZ61 magnesium alloy through a multitemperature multi-axial forging (MAF) process aimed at enhancing mechanical properties. By systematically reducing the processing temperature in successive forging passes, a fine-grained microstructure with an average grain size of 8 μm was achieved. The process promoted dynamic recrystallization, which refined the microstructure and improved strength, hardness, and ductility. The results showed that equivalent strain significantly influenced microstructural homogeneity and mechanical performance, but further grain refinement ceased once dynamic recrystallization was completed. This multitemperature MAF approach demonstrates a scalable and effective strategy for producing high-strength, fine-grained magnesium alloys suitable for industrial applications.

In a study by Fujikawa et al. [26], the benefits of FEM simulation in the aluminum casting-forging process were discussed. The study emphasizes that focusing on the elimination of small pores in cast products, rather than increasing the forging ratio, is more efficient. FEM analysis was used as a critical tool in this process, demonstrating that the plastic deformation of multi-holed objects can be applied to casting-forging operations. While aluminum casting-forging applications have not yet become widespread, the wide combinations of existing materials and processes show that this process has significant potential. Additionally, it has been demonstrated that the forging process can eliminate internal defects in the casting, producing materials with excellent mechanical properties. The simulation technique is expected to have more importance in the future development of new manufacturing processes.

In another study, the hybrid manufacturing method combining stir casting and hot forging processes has been used to produce composite materials [29]. In the first phase, metal matrix composites were prepared using the stir casting method and then shaped through hot forging. This hybrid method combines the advantages of casting and forging processes, providing better mechanical properties and ease of shaping. As a result, it was observed that materials produced using this hybrid method had a more homogeneous structure and improved mechanical properties.

Another study developed a simulation approach for the hybrid evaluation of casting and forging processes [30]. This hybrid method improves the surface quality and microstructure of parts by closing the pores and air gaps in casting through the forging process. Segregations and material heterogeneities in cast parts are corrected through forging, leading to a more homogeneous structure. The study emphasizes that by combining casting and forging simulations, material losses can be reduced, and more efficient production processes can be achieved. Furthermore, this method offers significant opportunities for producing customized, lighter, and more efficient designs based on part functions.

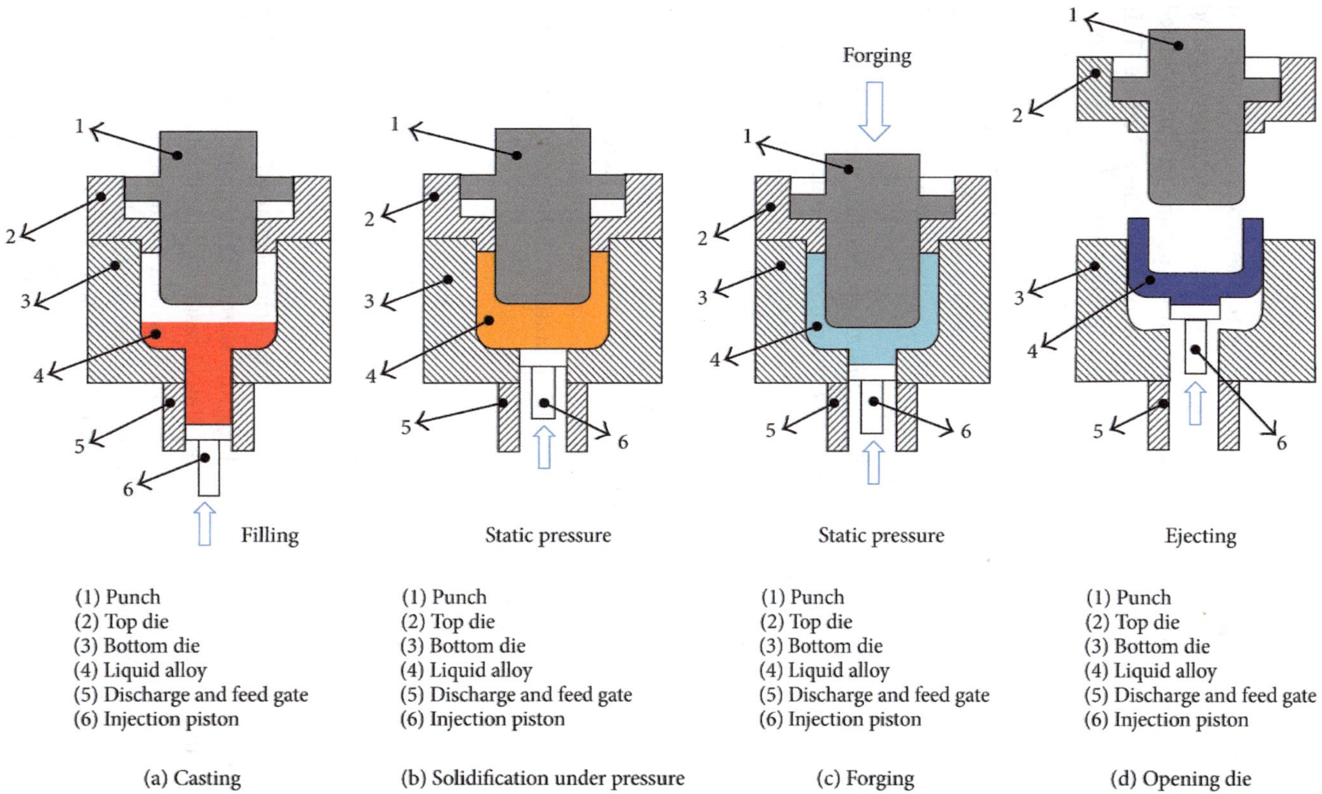

Fig. 5.4: Process principle of ICFP [27]. From *Advances in Mechanical Engineering* (Sage), used under Creative Commons Attribution (CC BY) 3.0 License.

5.2.5 Laser-Assisted Casting

Laser-assisted casting is a production process similar to laser coating methods but without bonding between the substrate and the coating [31]. This method is applied by placing laser-melted metal into a mold, where the process parameters are adjusted to intentionally prevent the molten metal from bonding with the substrate. This makes it possible to produce objects directly in cast form, independent of the substrate. Laser-assisted casting is an innovative technique, especially preferred in applications where bonding is not desired. During the process, factors such as the interaction of the molten pool with the powder bed, the advancement of molten metal, and surface roughness are controlled based on thermodynamic and fluid dynamics principles. Laser casting stands out as a production technique that allows the creation of complex shapes and eliminates bonding issues.

The interaction of powder particles with the molten pool during the laser casting process and their effects on surface quality were examined in detail [32]. The study found that smaller particles were absorbed more quickly by the molten pool, while larger and partially melted particles accumulated on the liquid surface, creating a rough structure. As a result, it was observed that the laser casting surface was not sufficiently wetted by the substrate, preventing the formation of strong bonds or smooth surfaces. Additionally, it was determined that process parameters could cause a transition between laser casting and laser coating. When appropriate parameters were not provided, the surface was covered with semi-melted protrusions, resulting in a rough structure. These findings highlight the limitations of the laser casting process in producing smooth surfaces.

5.2.6 Ultrasonic-Assisted Casting

5.2.6.1 Ultrasonic-Assisted Stir Casting

Ultrasonic-assisted stir casting is an advanced casting technology used in the production of metal matrix composite materials [33]. This method is an approach that improves the traditional stir casting process by using ultrasonic vibrations to ensure the homogeneous distribution of reinforcement materials within the metal matrix. Ultrasonic vibrations reduce the surface tension of the liquid metal, allowing better wetting and homogeneous distribution of the reinforcement materials. It also reduces porosity, facilitates gas removal, and improves grain structure, thereby enhancing the mechanical properties of composite materials. This method provides superiority, especially in the production of metal matrix nanocomposites containing nanoscale reinforcement materials and offers significant improvements in mechanical properties such as hardness and tensile strength [33, 34].

Ultrasonic-assisted stir casting has been successfully applied in the production of Al 2219 alloy metal matrix nanocomposites, providing significant advantages [35]. This method ensured the homogeneous distribution of nano-alumina (Al_2O_3) particles within the matrix, improving mechanical properties. Additionally, due to the grain refinement and degassing effects of ultrasonic vibrations, low porosity was achieved. With this method, hardness reached 260 HV, and tensile strength reached 250 MPa, representing a 32% increase in these values. These results demonstrate that ultrasonic-assisted stir casting is an effective method for producing nanocomposites with homogeneous reinforcement distribution and superior mechanical properties.

The production of Al-SiC (aluminum-silicon carbide) micro-composites using stir casting and ultrasonic-assisted stir casting methods was investigated, and the results were analyzed and compared [34]. X-ray diffraction analysis showed that Al and SiC were the main elements, and no new phase formation was observed. As the percentage of SiC increased, the intensity of the SiC peaks also increased. Microstructure analysis revealed that in traditional stir casting, SiC particles were not distributed homogeneously and tended to cluster, while in ultrasonic stir casting, SiC particles were homogeneously distributed within the metal matrix. The ultrasonic mixing effect helped break up SiC particle clusters, leading to improvements in the microstructure. Tensile strength test results showed that the sample with 10% SiC content in the ultrasonic-assisted stir casting method had a tensile strength 11.4% higher than the traditional stir casting method. In compression strength tests, a 7.9% improvement was achieved with ultrasonic-assisted stir casting. Hardness tests showed that with ultrasonic-assisted stir casting, hardness values reached 47.05%. These results indicate that ultrasonic mixing strengthens the interface bonds between SiC and the matrix and reduces voids. Density measurements showed that as SiC content increased, density increased by 0.63%, and ultrasonic-assisted stir casting provided higher density values than traditional stir casting. In conclusion, ultrasonic-assisted stir casting significantly improved the mechanical properties of Al-SiC micro-composites compared to traditional stir casting. These improvements are based on enhanced tensile strength, compression strength, and hardness. This method ensured the homogeneous distribution of SiC and reduced microstructural voids.

The methodology applied for ultrasonic-assisted casting is like a simple casting technique. The schematic in Fig. 5.5 represents an ultrasonic-assisted stir casting process used to incorporate nanoparticles into a molten alloy. High-frequency vibrations generated by a transducer, improve nanoparticle dispersion and prevent agglomeration. The process operates under controlled temperature and an inert argon atmosphere to ensure uniform distribution and minimize oxidation during composite fabrication.

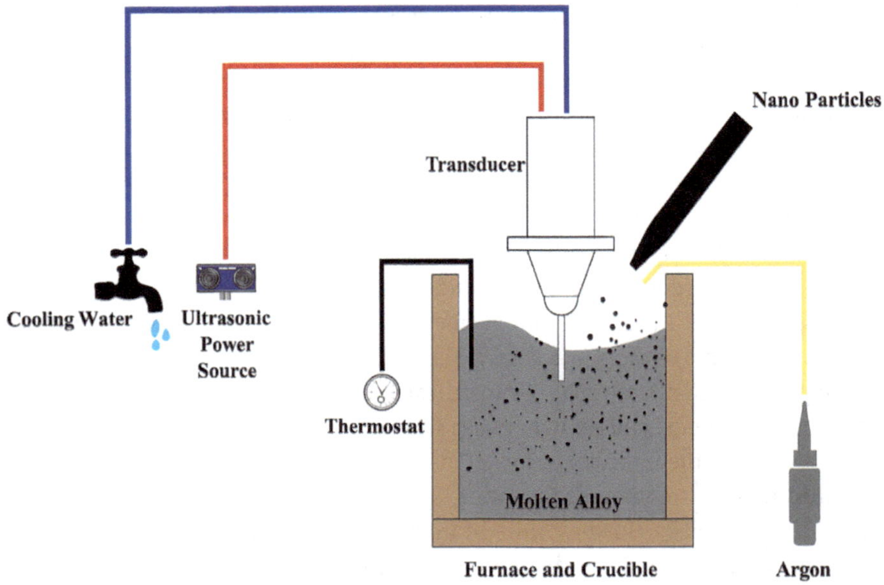

Fig. 5.5: Ultrasonic-assisted stir casting process.

5.2.6.2 Ultrasonic-Assisted Squeeze Casting

Squeeze casting is a solidification processing technique in which molten metal is directly solidified under high pressure within a closed die, combining features of both casting and forging. This method enhances mold filling and solidification control, offering advantages such as reduced shrinkage porosity, increased cooling rates, and refined grain structures, all of which contribute to improved mechanical properties of the alloy [36]. However, squeeze casting can lead to a bimodal structure, causing a simultaneous distribution of large and small grains, while ultrasonic treatment exhibits a damping effect over longer distances in the liquid. In recent years, the combination of ultrasonic treatment and squeeze casting has emerged as a more effective method for improving the properties of casting alloys.

The application of ultrasonic treatment during solidification has emerged as an effective and environmentally friendly method for refining grain size, homogenizing the chemical composition, reducing shrinkage porosity, and improving the mechanical properties of alloys [36]. It is known that ultrasonic treatment during the solidification process relies on two main mechanisms: acoustic cavitation, which breaks dendrites by the formation, growth, and collapse of bubbles in the liquid, thereby increasing nucleation rates, and acoustic flow, which promotes the chemical homogeneity of alloys by moving the molten metal. The illustration of working principle of ultrasonic-assisted squeeze casting is shown in Fig. 5.6. Molten alloy is poured into a cavity die, and pressure is applied via a punch to facilitate densification during solidification. Simultaneously, an ultrasonic probe introduces high-frequency vibrations into the alloy melt to improve melt flow, break up gas bubbles, and refine the microstructure. A thermocouple is inserted to monitor the temperature throughout the process, while the sample forms under the combined effects of pressure and ultrasonic energy.

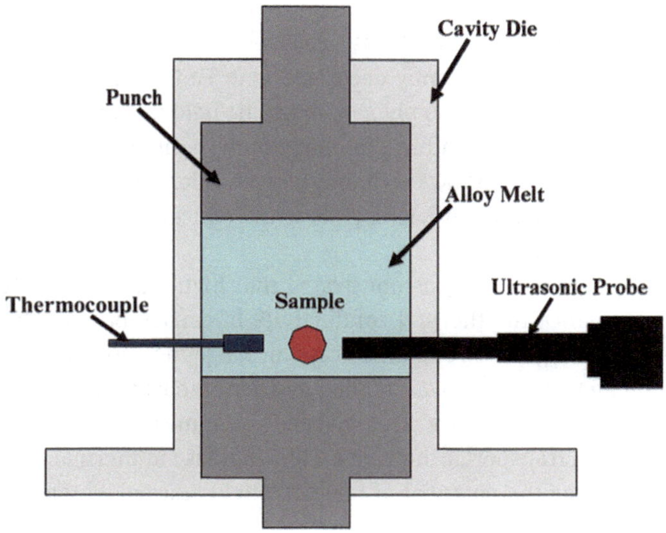

Fig. 5.6: Schematic diagram of ultrasonic-assisted squeeze casting.

5.2.7 Magnetic-Field-Assisted Casting

Magnetic-field-assisted casting (MAC) is a method that uses a magnetic field during the casting process to align particles in a specific arrangement [37]. Especially when magnetic materials or magnetic nanoparticles are used as fillers, the magnetic field helps to control the orientation of these particles within the liquid matrix. This method prevents the random distribution of particles, allowing the material's mechanical, thermal, and electrical properties to be improved in specific directions.

In a study, the advantages of the rotating magnetic-field-assisted casting (RMAC) method in creating bidirectional thermal conductivity pathways were investigated [38]. The RMAC method uses a rotating magnetic field to control the alignment of magnetic functional graphene (m-FG) particles within the liquid matrix. This allows the particles to align in a specific direction, rather than randomly distributing, significantly improving the thermal conductivity performance of the material. Particularly, the creation of bidirectional thermal conductivity pathways offers significant advantages for applications that require high thermal performance by optimizing heat transfer. Compared to randomly aligned composites, the composites produced using the RMAC method exhibited a more organized particle arrangement, thus showing more consistent and superior performance in both thermal and mechanical properties. As a result, the RMAC method offers an important solution for advancing the thermal and physical performance of composite materials in engineering applications.

Ryu et al. [37] present a study on enhancing the through-plane thermal conductivity of epoxy-based thermal interface materials (TIMs) using magnetically aligned electrochemically exfoliated graphite (EEG). The EEG was selectively oxidized at the edges (sGO) to improve dispersibility and subsequently decorated with Fe_3O_4 nanoparticles (msGO) to enable magnetic alignment. Upon applying a magnetic field during the curing process, the msGO aligned vertically within the epoxy matrix, significantly improving thermal conductivity to 1.10 W m^{-1} K^{-1} with only 25 wt% filler. This approach offers an effective strategy for designing efficient TIMs with reduced filler content and enhanced heat dissipation performance.

Finally, in their study, Chung et al. [39] demonstrated that high magnetic fields can be effectively used to enhance the thermal conductivity of carbon-based TIMs through filler alignment. Exploiting the strong anisotropy in magnetic susceptibility of graphite, a 10 T superconducting magnet was applied to orient the c-axis of graphite platelets perpendicular to the film surface. This magnetic alignment significantly improved through-plane thermal transport, achieving a 330% increase in thermal conductivity. Their findings underline the potential of magnetic field-assisted alignment for tailoring microstructure and anisotropic heat conduction in composite materials.

5.2.8 Electromagnetic Casting

Electromagnetic casting (EAC) is a hybrid manufacturing method that combines traditional casting techniques with electromagnetic stirring (EMS) applications [40]. This method is based on the internal electromagnetic stirring – direct chill casting (IEMS-DCC) method to improve the homogeneity of liquid metal and eliminate microstructural deficiencies.

EAC improves the microstructure of liquid metal, resulting in finer and more homogeneous crystalline structures. By combining EMS with intercooling, the temperature distribution during cooling becomes more balanced. Some of the main advantages of this method include microstructure homogeneity, improved mechanical properties, and reduced material defects during casting [40, 41]. The homogeneous microstructures obtained through this method are particularly suitable for high-performance applications in industries such as aerospace, automotive, and aviation. EAC was developed to overcome the microstructural deficiencies and performance limitations of traditional casting methods, while also providing better machinability and strength properties. Therefore, it holds significant potential for future hybrid manufacturing methods.

Qiu et al. [40] investigated the effect of IEMS during direct chill casting on the hot workability of 2219 aluminum alloy. By placing the electromagnetic stirrer inside the melt, IEMS generated a more uniform thermal field and refined grain structure compared to conventional methods. This magnetic approach enhanced the processing map by expanding the stability domain, particularly allowing effective deformation at lower temperatures and higher strain rates. The IEMS-DCC billets exhibited improved microstructural homogeneity and promoted dynamic recrystallization, demonstrating that EAC techniques can significantly improve deformation behavior through electromagnetic field–induced control of solidification.

Low-frequency electromagnetic casting (LFEC) is an advanced method used in liquid metal casting that utilizes low-frequency electromagnetic fields to control the movement of the liquid metal [42]. The electromagnetic forces applied to the liquid metal enhance its homogeneity, reducing micro-segregation and optimizing the mechanical properties of alloys. LFEC allows for more precise control over the heterogeneous structures and grains in the liquid metal. A LFEC setup, as shown in Fig. 5.7, where a coil generates a magnetic field to stir the alloy melt. The system includes a hot top, crystallizer mold, graphite ring, and water cooling. EMS refines the microstructure by enhancing melt flow and promoting uniform solidification in the ingot.

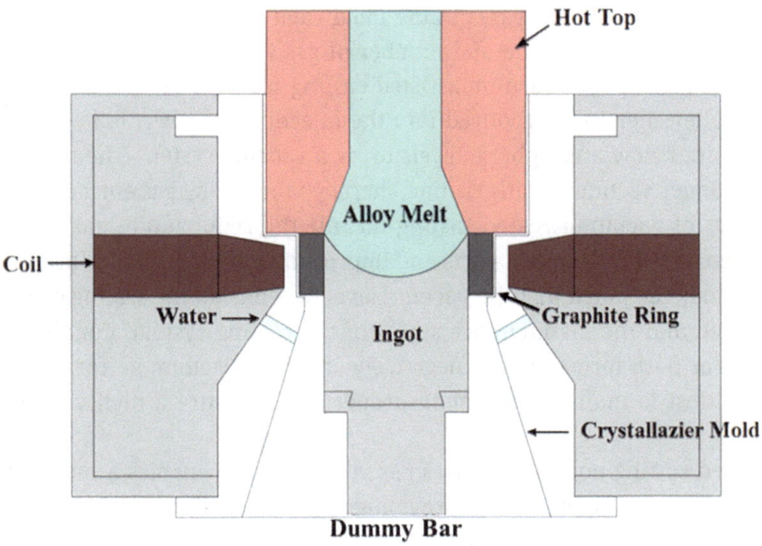

Fig. 5.7: Schematic diagram of low-frequency electromagnetic casting experiment.

Wang et al. [42] investigated the influence of LFEC on 5A90 aluminum-lithium alloys, demonstrating that applying a 10 Hz/100 A electromagnetic field during casting significantly refined the microstructure, reduced macro-segregation, and improved mechanical properties. The Lorentz force induced forced melt flow, which enhanced heat transfer and homogenized the solidification front, leading to finer equiaxed grains and suppressed eutectic formation. Compared to conventional direct chill casting, LFEC enhanced yield strength, reduced grain boundary segregation, and increased yield rate by 2.3%, making it an effective approach for structural quality improvement in aluminum alloy ingots.

5.2.9 Vacuum-Assisted Casting

High-pressure casting is a method with a wide range of applications in industrial production. In this process, molten metal is injected into the mold at high speeds [43, 44]. Some issues can arise in the high-pressure casting method. One of the main problems is the formation of gas voids [43]. The entrapment of air or gases inside the mold can lead to a porous structure in the cast parts and a reduction in mechanical properties. The primary factors contributing to the formation of gas voids include trapped air in the injection chamber and mold cavity, gases resulting from the burning or evaporation of plunger lubricants, and dissolved hydrogen in the metal. Figure 5.8 illustrates vacuum-assisted die casting, where a vacuum pump removes air from the die cavity before molten metal is injected by a plunger, enhancing filling quality and reducing porosity.

Vacuum-assisted pressure casting is a method developed to overcome these issues [44, 45]. In this process, the air and gases in the mold cavity are evacuated with the help of a vacuum. As a result, the size and number of gas voids are significantly reduced. The working principle of vacuum-assisted casting involves closing the mold cavity after the molten metal is introduced into the injection chamber, followed by reducing the pressure below atmospheric levels using a vacuum system. The mold is continuously kept under vacuum during casting, thereby minimizing gas entrapment.

The advantages of vacuum-assisted casting include the reduction of gas voids, leading to increased density of the cast parts and improvement in mechanical properties. However, it should be noted that the vacuum level depends on the sealing condition inside the mold and the maintenance status of the casting system. Conditions such as mold leaks or flash formation can negatively affect the vacuum performance. Therefore, it is critical to maintain the mold properly and ensure a high vacuum level.

Vacuum-assisted casting not only reduces gas voids but also enables a more homogeneous microstructure. Significant improvements in the hardness and strength properties of the alloy are observed. This method has become an indispensable tool to produce high-quality cast parts. In a study, tensile test samples according to the American Society for Testing and Materials standards were produced from three different Al-Si-based alloys containing 5%, 8%, and 18% silicon using the vacuum-assisted high-pressure casting method [45]. It was found that vacuum-assisted casting reduced the amount of air and gases trapped in the mold cavity, significantly decreasing the volume and size of gas voids. These improvements increased the mechanical properties and density of the cast parts. Furthermore, it was determined that the injection speed

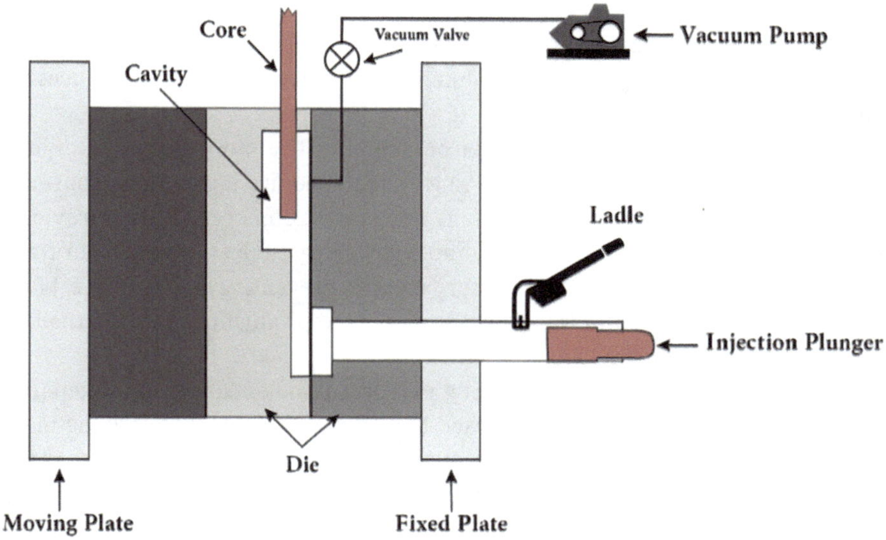

Fig. 5.8: The principle of a vacuum-assisted die casting process.

had a significant effect on mechanical properties, with higher injection speeds providing optimal characteristics. After the T6 heat treatment, it was observed that vacuum-assisted casting reduced the bubbles formed on the surface and further improved the mechanical properties of the heat-treated alloys. These results demonstrate that parts produced by vacuum-assisted casting are suitable for heat treatment.

5.3 Environmental Impacts and Ethical Evaluations in Casting-Based Hybrid Manufacturing

5.3.1 Sustainability and Environmental Impacts

Casting-based hybrid manufacturing makes significant contributions to sustainability principles in terms of environmental impacts and ethical evaluations. Sustainability forms the basis of these manufacturing processes, while evaluating the environmental impacts of hybrid manufacturing technologies is of great importance. These processes aim to reduce environmental impacts by providing innovative approaches in areas such as material efficiency, energy use, recycling practices, and carbon footprint.

Material efficiency is one of the prominent advantages of casting-based hybrid manufacturing systems. While material loss can be high in traditional casting processes, hybrid technologies minimize material waste through methods such as AM. These methods reduce waste by using only the necessary amount of material, offering sustainable material management during production. Additionally, the use of recycled raw materials helps mitigate environmental impacts and reduce production costs. The use of recyclable materials such as aluminum and steel significantly reduces the carbon footprint.

Energy efficiency is another environmentally friendly aspect of hybrid manufacturing processes. Traditional casting systems require high energy consumption, while hybrid manufacturing technologies optimize energy use by providing more precise heat management. AM technologies consume less energy than traditional methods by only heating the necessary regions, preventing energy losses. Moreover, the technologies used in these processes reduce production time, further decreasing energy consumption [46].

The zero-waste approach and recycling practices also contribute to the environmental sustainability goals of casting-based hybrid manufacturing. Recycling metal waste generated during these processes optimizes resource use and minimizes environmental impacts. For example, integrating metals like aluminum into the circular economy is an effective method for reducing carbon emissions in production processes. Hybrid systems support zero-waste goals by offering innovative solutions in waste management.

Furthermore, hybrid manufacturing processes have the potential to reduce carbon footprints and emissions. More efficient energy use and the integration of renewable energy sources can significantly reduce the environmental impacts of these systems. Replacing fossil fuel use with renewable energy sources helps lower carbon emissions. However, the energy consumption of hybrid systems should still be carefully assessed, as high-temperature processes like casting are energy-intensive.

Finally, the environmental impacts of materials used in hybrid manufacturing should also be considered in evaluating the process. The selection of high-performance and environmentally friendly materials can enhance the environmental benefits of hybrid manufacturing processes. Biodegradable materials and eco-friendly binders support the environmental sustainability of these processes and open the door for new material innovations. Thus, casting-based hybrid manufacturing systems offer a production process that aligns with environmental and ethical responsibilities.

5.3.2 Ethical Evaluations

The ethical evaluation of hybrid manufacturing technologies goes beyond environmental impacts and encompasses a broad range of issues such as workforce, education, public health, and social balance. The increasing use of automation systems has led to fundamental changes in production, with significant consequences for the workforce. Robots and automated processes replace human labor on the production line, potentially increasing the risk of unemployment. However, the widespread adoption of hybrid manufacturing technologies can also create new job opportunities by increasing the demand for skilled labor in fields such as engineering, design, and technology management. In this transformation process, retraining and adaptation of the workforce is critical. If education systems are unable to keep up with this technological change, especially in developing regions, social inequalities are likely to deepen.

The impact of hybrid manufacturing systems is not limited to the workforce; the high technical knowledge and skills required by these technologies also necessitate the restructuring of educational policies. However, this situation can further highlight educational inequalities in regions with limited access to technology. The widespread use of high-tech hybrid systems can both enable the development of related programs in technical schools and universities and create serious problems regarding the accessibility of these programs. Ensuring the adaptation of societies to these new manufacturing processes is possible not only through the widespread dissemination of technical education but also by making these education opportunities accessible to everyone.

Environmental and societal impacts are also key dimensions of the evaluation of hybrid manufacturing technologies. The chemicals used in casting-based hybrid manufacturing processes can potentially harm the environment and public health, requiring careful management of these processes. Similarly, emissions and energy consumption during production can have long-term negative effects on both the environment and society. Therefore, it is crucial that hybrid manufacturing technologies are designed and implemented in a way that supports environmental sustainability. The use of automation and robotic technologies can offer advantages for workforce safety; however, if these systems are not properly monitored, new risks may emerge.

The societal impacts of hybrid manufacturing technologies are particularly evident in developing regions where inequalities in access to technology exist. The rapid adoption of these technologies in developed countries can negatively affect the competitiveness of producers in developing countries. Moreover, the financial and technical barriers faced by small and medium-sized enterprises (SMEs) during the transition to high-tech systems can lead to a loss of market share for these companies. While large companies can more easily integrate such technologies, this can increase pressure on local economies. On a global scale, the widespread adoption of hybrid manufacturing systems risks deepening inequality in access to technology. To prevent this, inclusive policies must be developed, and more equitable support mechanisms for transitioning to hybrid manufacturing technologies should be established.

In conclusion, the ethical dimensions of hybrid manufacturing technologies should be carefully addressed to balance both their benefits and potential negative impacts. Expanding educational opportunities, adopting environmental sustainability goals, and increasing support for local producers can ensure that hybrid manufacturing systems are developed in a way that maximizes social benefits. This will not only reduce existing inequalities but also contribute to a more equitable industrial transformation.

Acronyms and Nomenclature

3D	Three dimension
3DSP	3D sand printing
ABS	Acrylonitrile-butadiene-styrene
ADI	Austempered ductile ıron
Al	Aluminum
Al_2O_3	Alumina (aluminum oxide)
AM	Additive manufacturing
Ar	Argon
C	Carbon
CAD	Computer-aided design
CCF	Conventional cutting fluid
CHSC	Conical-helix sprue
CS	Cast steel
Cu	Copper
DC	Direct chill
DCC	Direct chill casting
DLMS	Direct laser metal sintering
EAC	Electromagnetic casting
EEG	Electrochemically exfoliated graphite
EMS	Electromagnetic stirring
FDM	Fused deposition modeling
Fe	Iron
Fe_3O_4	Magnetite (iron oxide) nanoparticles
FEM	Finite element method
FG	Functional graphene
FSW	Friction stir welding
HF	Hydrofluoric acid
HIC	Hybrid ınvestment casting
HV	Vickers hardness
IC	Investment casting
ICFP	Integrated casting and forging process.
IEMS	Internal electromagnetic stirring.
IEMS-DCC	Internal electromagnetic stirring – direct chill casting.
Inconel	Nickel-based superalloy
LCF	Low cycle fatigue
LDP	Laser powder bonding
LFEC	Low-frequency electromagnetic casting
Li	Lithium
L-PBF	Laser powder bed fusion
MAC	Magnetic-field-assisted casting
MAF	Multi-axial forging
MAGMASOFT	Casting simulation software
m-FG	Magnetic functional graphene
Mg	Magnesium
Mo	Molybdenum
MPa	Megapascal
MQL	Minimum quantity lubrication

msGO	Fe_3O_4-decorated sGO
Ni	Nickel
PWHT	Post-weld heat treatment
R&D	Research and development
RMAC	Rotating magnetic-field-assisted casting
RP	Rapid prototyping
SEM	Scanning electron microscopy/micrographs
sGO	Selectively oxidized graphene oxide
Si	Silicon
SiC	Silicon carbide
SLA	Stereolithography
SLM	Selective laser melting
SLS	Selective laser sintering
Sn	Tin
T	Tesla (magnetic flux density)
T6	Heat-treatment temper designation
TIM	Thermal interface material
Ti	Titanium
UTS	Ultimate tensile strength
V	Vanadium
V/A	Volume-to-area ratio
$W\ m^{-1}\ K^{-1}$	Thermal conductivity unit (Watts per meter per Kelvin)
Zn	Zinc
Zr	Zirconium
α	Alpha phase (hexagonal close-packed microstructure)
β	Beta phase (body-centered cubic microstructure)

References

[1] Shah M, R. Patel D, Pande S. Additive Manufacturing Integrated Casting- A Review, In: Materials Today. Elsevier Ltd; 2022. 62, p. 7199–7203. https://doi.org/10.1016/j.matpr.2022.03.413.

[2] Kumar P, Singh R, Ahuja I. A Framework for Developing a Hybrid Investment Casting Process, Asian Rev Mech Eng. 2013;2:49–55.

[3] Sebbe NPV, Fernandes F, Sousa VFC, Silva FJG. Hybrid Manufacturing Processes Used in the Production of Complex Parts: A Comprehensive Review, Metals. 2022;12. https://doi.org/10.3390/met12111874.

[4] Lauwers B, Klocke F, Klink A, Tekkaya AE, Neugebauer R, McIntosh D. Hybrid Processes in Manufacturing, CIRP Ann Manuf Technol. 2014;63:561–583. https://doi.org/10.1016/j.cirp.2014.05.003.

[5] Ferro P. Casting and Solidification Processing, Metals. 2022;12. https://doi.org/10.3390/met12040559.

[6] Iraola-Arregui I, Ben Youcef H, Trabadelo V. Cost estimation tool for metallic parts made by casting: A case study, Metals. 2023;13. https://doi.org/10.3390/met13020216.

[7] Dias LRM, Diniz AE. Effect of the gray cast iron microstructure on milling tool life and cutting force, J Braz Soc Mech Sci Eng. 2013;35:17–29. https://doi.org/10.1007/s40430-013-0004-3.

[8] Eraslan D, Balcı A, Çetin B, Uçak N, Çiçek A, Yılmaz OD. et al. Machinability evaluations of austempered ductile iron and cast steel with similar mechanical properties under eco-friendly milling conditions, J Mater Res Technol. 2021;11:1443–1456. https://doi.org/10.1016/j.jmrt.2021.01.123.

[9] Sefer B, Dobryden I, Almqvist N, Pederson R, Antti ML. Chemical milling of cast Ti-6Al-4V and Ti-6Al-2Sn-4Zr-2Mo alloys in hydrofluoric-nitric acid solutions, Corrosion. 2017;73:394–407. https://doi.org/10.5006/2277.

[10] Fernández-Calvo AI, Madarieta M, Solana A, Lizarralde I, Rouco M, Soriano C. Study of the hibridation of ablation casting and laser wire metal deposition for aluminum alloy 5356, Crystals. 2025;15. https://doi.org/10.3390/cryst15020134.

[11] Zong X, Ji Z, Zhao Z, Feng X, Ding X, Nan H. Microstructure and tensile property of hybrid fabricated Ti-6Al-4V alloy by investment casting and laser additive manufacturing, Metals. 2023;13. https://doi.org/10.3390/met13040668.

[12] Marques ESV, Silva FJG, Paiva OC, Pereira AB. Improving the mechanical strength of ductile cast iron welded joints using different heat treatments, Materials. 2019;12. https://doi.org/10.3390/ma12142263.

[13] Senff M, Volk W. Hybrid joining of cast aluminum and steel by compound casting and friction stir welding, Prod Eng. 2023;17:521–534. https://doi.org/10.1007/s11740-023-01184-9.

[14] Gouveia RM, Silva FJG, Paiva OC, De fátima andrade M, Pereira LA, Moselli PC. et al. Comparing the structure and mechanical properties of welds on ductile cast iron (700 MPa) under different heat treatment conditions, Metals. 2018;8: https://doi.org/10.3390/met8010072.

[15] Wu KJ, Xian MQ. The role and impact of 3D printing technologies in casting, China Foundry. 2017;14:157–168. https://doi.org/10.1007/s41230-017-6109-z.

[16] Liou FF. Rapid Prototyping and Engineering Applications. 2nd ed, Taylor & Francis, CRC Press; 2019. https://doi.org/10.1201/9780429029721.

[17] Xie S, Lv Z, Dong S. Study on the optimization of investment casting process of exhaust elbow for high-power engine, Metals. 2024;14. https://doi.org/10.3390/met14040481.

[18] Kumar P, Ahuja IS, Singh R. Effect of process parameters on surface roughness of hybrid investment casting, Prog Addit Manuf. 2016;1:45–53. https://doi.org/10.1007/s40964-016-0004-9.

[19] Dave HK, Dixit US, Nedelcu D. editors. Recent Advances in Manufacturing Processes and Systems. 1st ed, Singapore: Springer Nature Singapore; 2022. https://doi.org/10.1007/978-981-16-7787-8.

[20] Wu M, Tinschert J, Augthun M, Wagner I, Scha Èdlich-Stubenrauch J, Sahm PR. et al. Application of laser measuring, numerical simulation and rapid prototyping to titanium dental castings, Dent Mater. 2001;17(2):102–108.

[21] Lynch P, Hasbrouck C, Wilck J, Kay M, Manogharan G. Challenges and opportunities to integrate the oldest and newest manufacturing processes: Metal casting and additive manufacturing, Rapid Prototyp J. 2020;26:1145–1154. https://doi.org/10.1108/RPJ-10-2019-0277.

[22] Sama SR, Badamo T, Lynch P, Manogharan G. Novel sprue designs in metal casting via 3D sand-printing, Addit Manuf. 2019;25:563–578. https://doi.org/10.1016/j.addma.2018.12.009.

[23] Almaghariz ES, Conner BP, Lenner L, Gullapalli R, Manogharan GP, Lamoncha B. et al. Quantifying the role of part design complexity in using 3D sand printing for molds and cores, Int J Metalcasting. 2016;10:240–252. https://doi.org/10.1007/s40962-016-0027-5.

[24] Krochmal M, Rajan ANR, Moeini G, Sajadifar SV, Wegener T, Niendorf T. Microstructural and mechanical properties of AlSi10Mg: Hybrid welding of additively manufactured and cast parts, J Mater Res. 2023;38:297–311. https://doi.org/10.1557/s43578-022-00838-1.

[25] Arbo SM, Tjøtta S, Boge MH, Tundal U, Li J, Dumoulin S, Jensrud O. The potential of cast stock for the forging of aluminum components within the automotive industry, Metals. 2024;14. https://doi.org/10.3390/met14010090.

[26] Fujikawa S, Kitamura Y, Shimamura S. Application of numerical methods for the aluminum casting/forging process, Journal of Materials Processing Technology. 1991;27.

[27] Zhang Q, Cao M, Zhang D, Zhang S, Sun J. Research on integrated casting and forging process of aluminum automobile wheel, Adv Mech Eng. 2014; 2014 https://doi.org/10.1155/2014/870182.

[28] Chen Q, Shu D, Hu C, Zhao Z, Yuan B. Grain refinement in an As-Cast AZ61 magnesium alloy processed by multi-axial forging under the multitemperature processing procedure, Mater Sci Eng. 2012;541:98–104. https://doi.org/10.1016/j.msea.2012.02.009.

[29] Shivananda Murthy KV, Girish DP, Keshavamurthy R, Varol T, Koppad PG. Mechanical and thermal properties of AA7075/TiO2/Fly ash hybrid composites obtained by hot forging, Prog Nat Sci Mater Int. 2017;27:474–481. https://doi.org/10.1016/j.pnsc.2017.08.005.

[30] Böhmichen U, Schubert N, Lehnert T, Sterzing A, Mauermann R. From casting to forging – the combined simulation for a steel component, Eng Rep. 2022;4. https://doi.org/10.1002/eng2.12400.

[31] Powell J, Gedda H, Kaplan A. Laser casting and laser clad-casting: New processes for rapid prototyping and production, In: 21st International Congress on Applications of Laser and Electro-Optics, Congress Proceedings. Laser Institute of America; 2002. https://doi.org/10.2351/1.5066171.

[32] Gedda H, Kaplan A, Powell J. Melt-Solid interactions in laser cladding and laser casting, Metall Mater Trans B. 2005;36B:683–689.

[33] Nie KB, Wang XJ, Wu K, Xu L, Zheng MY, Processing HXS. Microstructure and mechanical properties of magnesium matrix nanocomposites fabricated by semisolid stirring assisted Ultrasonic Vibration B, J Alloys Compd. 2011;509:8664–8669. https://doi.org/10.1016/j.jallcom.2011.06.091.

[34] Idrisi AH, Mourad AHI. Conventional stir casting versus ultrasonic assisted stir casting process: Mechanical and physical characteristics of AMCs, J Alloys Compd. 2019;805:502–508. https://doi.org/10.1016/j.jallcom.2019.07.076.

[35] Murthy NV, Reddy AP, Selvaraj N, Rao CSP. Dispersion of Alumina nano particles in Al 2219 alloy by ultrasonic assisted stir casting technique, Mater Today Proc. 2017;4:10113–10117.

[36] Zhao Y, He W, Yang Y, Liu H, Wei Q, Lin B. et al. Effect of ultrasonic treatment and squeeze casting on the microstructural refinement of Al–Cu–Mn alloys, Int J Metalcasting. 2024;18:869–881. https://doi.org/10.1007/s40962-023-01092-9.

[37] Ryu JH, Yang SM, Lee JU, Kim JH, Yang SJ. Magnetic alignment of electrochemically exfoliated graphite in epoxy as a thermal interface material with high through-plane thermal conductivity, Carbon Lett. 2022;32:1433–1439. https://doi.org/10.1007/s42823-022-00378-y.

[38] Zhao Z, Xiong R, Deng T, Zhang L, Li C. Dual-Direction thermal conductive fg-based composite manufactured by rotating magnetic-field assisted casting, Compos Sci Technol. 2024;247. https://doi.org/10.1016/j.compscitech.2023.110426.

[39] Chung SH, Kim H, Jeong SW. Improved thermal conductivity of carbon-based thermal interface materials by high-magnetic-field alignment, Carbon. 2018;140:24–29. https://doi.org/10.1016/j.carbon.2018.08.029.

[40] Qiu Y, Zheng K, Li X, Luo Y, Xia P, Liu M. et al. Processing map of 2219 Al alloy prepared by internal electromagnetic stirring direct chill casting, J Mater Res Technol. 2022;18:2885–2895. https://doi.org/10.1016/j.jmrt.2022.03.177.

[41] Qiu Y, Zhang Z. Effects of internal electromagnetic stirring on the microstructure refinement and composition homogenization of large-scale 7075 Aluminum alloy billet, JOM. 2021;73:3812–3818. https://doi.org/10.1007/s11837-021-04875-9.

[42] Wang F, Wang X, Cui J. Effect of low-frequency electromagnetic casting on micro-structure and macro-segregation of 5A90 alloy ingots, Materials. 2020;13. https://doi.org/10.3390/ma13122720.

[43] Street A. The Die Casting Book. 1st ed, Portcullis Press Ltd; 1977.

[44] Niu XP, Hu BH, Pinwill I, Li H. Vacuum assisted high pressure die casting of Aluminium alloys, J Mater Process Technol. 7 Sep 2000; 105(1-2): 119–127.
[45] Wang QL, Xiong SM. Vacuum assisted high-pressure die casting of AZ91D Magnesium alloy at different slow shot speeds, Trans Nonferrous Met Soc China. 2014;24:3051–3059. https://doi.org/10.1016/S1003-6326(14)63442-7.
[46] Salonitis K, Zeng B, Mehrabi HA, Jolly M. The challenges for energy efficient casting processes, In Procedia CIRP. Elsevier B.V.; 2016. 40, p. 24–29. https://doi.org/10.1016/j.procir.2016.01.043.

Orhan Çakır, Çağla Özbilek, Beyzanur Yanık, Miraç Mutlu, Salih Yeşilsancak, Mohammad Zaher Akkad

6 Forming-Based Hybrid Manufacturing Processes

Abstract: Forming-based hybrid manufacturing represents a strategic integration of traditional forming techniques with auxiliary processes to overcome limitations such as restricted geometries, tool wear, and material formability. Classical methods like forging, deep drawing, extrusion, and incremental forming remain foundational in producing high-strength components with efficient material usage. However, challenges such as high tooling costs, limited flexibility, and springback persist. Hybrid manufacturing addresses these challenges by combining forming with processes including machining, thermal assistance, vibration, or surface treatments, either simultaneously or sequentially. This chapter provides a comprehensive overview of forming-based hybrid manufacturing processes, classifying them into combined, assisted, and sequential types. Combined approaches integrate multiple forming operations, while assisted methods enhance forming through external stimuli like heat or vibration. Sequential strategies coordinate forming with post-processes such as machining or hardening. Applications are illustrated through reviewing and discussing industrial and research case studies involving hybrid incremental forming, draw-forging, extrusion-bending, and hot stamping. These hybrid approaches not only improve dimensional accuracy and surface quality but also extend formability limits and reduce defects. Emphasis is placed on process mechanisms, performance benefits, and material behavior. This chapter underlines the transformative potential of hybrid forming in producing complex, high-performance, and cost-effective components, supporting advanced manufacturing needs.

Keywords: bending, extrusion, forming, hybrid manufacturing, plastic deformation, microstructure

6.1 Introduction

Traditional forming processes such as forging, stamping, and sheet metal forming have long been cornerstones of manufacturing [1, 2]. These "transformative" processes excel at efficiently shaping materials into high-strength components with near-net shapes, often with minimal waste compared to subtractive manufacturing processes like machining. For example, forging enables the production of dense and durable parts at a low cost per unit when scaled to high-volume manufacturing, making it a highly efficient and economically favorable process. Stamping and drawing operations enable

 | https://doi.org/10.1515/9783111704456-006

rapid production of complex sheet-metal geometries for automotive and aerospace structures. Beyond throughput, forming imparts desirable material properties (e.g., work-hardening or refined grain structure in forged parts) that enhance performance.

However, classic forming methods also face inherent limitations. They typically require dedicated dies or molds, leading to high tooling costs and inflexibility for low-volume or highly complex designs. Certain shapes and fine features remain impractical to form directly due to geometric constraints or lack of internal access. Moreover, material-related issues can arise: sheet metal processes often suffer from defects like wrinkling, excessive thinning, or springback (elastic shape recovery) that complicate dimensional control [2]. The formability of advanced high-strength alloys is limited, meaning stronger materials tend to crack or springback if pushed beyond modest strains. Overall, no single conventional forming technique can economically realize every geometry or precision requirement, and each has a "technological constraint" that leaves some component designs out of reach [3].

To address these limitations, the concept of hybrid manufacturing (HM) has emerged. HM involves combining two or more processes in a unified system or sequence to harness their complementary advantages. In essence, integrating multiple processes can create a synergistic effect greater than the sum of the individual methods (often described as a "1 + 1 = 3" outcome) [4]. According to Sebbe et al. [5], the objective of HM is the joining of two or more distinct processes in a single piece of equipment, observing the unique advantages of each one, while minimizing the limitations. Unlike a traditional manufacturing line where forming might be separate from other treatments, a hybrid approach intentionally overlaps or closely couples processes. This integration allows, for example, real-time adjustments or simultaneous effects that are impossible in isolation. Forming-based hybrid processes are specifically cantered on a metal forming operation enhanced by auxiliary processes such as machining, thermal energy input, surface treatment, or online measurement. By not relying solely on the forming die to achieve final shape and properties, such hybrid processes can overcome the shortcomings of conventional forming. It can be said that forming-based HM leverages the strengths of classical forming techniques while mitigating their weaknesses through strategic integration of auxiliary processes. By combining deformation with thermal, mechanical, or feedback control methods, these hybrid processes achieve improved formability, higher precision, extended tool life, and enhanced defect control [5, 6]. This chapter lays the groundwork for a comprehensive review of such hybrid processes. The subsequent sections will examine various implementations of forming-based HM, discuss their performance benefits over conventional methods, and highlight how they are expanding the capabilities of metal forming technology.

6.2 Forming Process

Forming processes involve shaping materials through plastic deformation under applied forces, as opposed to removing material (machining) or solidifying molten material (casting). These methods exploit the ductility of materials (especially metals) to achieve desired shapes while often improving or preserving material properties. Plastic forming encompasses a range of bulk forming processes (deforming the entire volume, typically with low surface-area-to-volume ratio) and sheet forming processes (deforming metal sheets with high surface-area-to-volume ratio). A classification of forming processes is shown in Fig. 6.1. This section presents the major plastic forming processes detailing their mechanisms, typical applications, and key technical considerations.

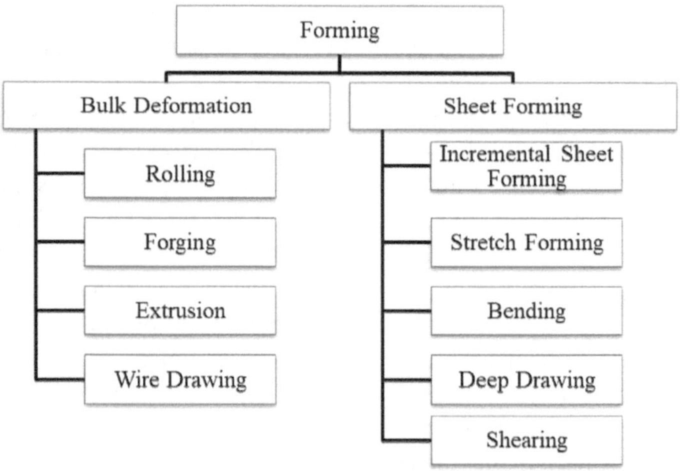

Fig. 6.1: Classification of forming process.

6.2.1 Forging

Forging is a deformation process in which metal is shaped by compressive forces, typically delivered through a hammer or press [7]. It is one of the oldest metal forming techniques, continuously improved to tackle modern requirements for complex shapes and high-performance parts. In forging, a heated billet (or sometimes a cold metal for small parts) is pressed between dies, causing plastic flow of the material into the desired shape. This process refines the metal's internal grain structure and often improves mechanical properties such as strength and toughness [7]. There are two major types of forging: open-die forging (free forging) and closed-die (impression) forging. Open-die forging involves deforming the workpiece between flat or simple-shaped dies; it is generally used for large, simple-shaped components and allows the production of parts with large dimensions and excellent mechanical properties [8]. It is typically employed for shafts, rings, disks, and other simple shapes (e.g., large turbine generator shafts and pressure vessel components). Closed-die forging uses shaped dies to confine flow, enabling near-net-shape parts with more complex geometry (common examples include crankshafts, connecting rods, and automotive gears). The forging process can be performed hot or cold. Hot forging (at high temperatures) reduces flow stress and facilitates large deformations but requires careful control of temperature and strain rate to avoid defects [9]. Key process parameters in forging include the workpiece temperature, strain rate (or ram speed), die geometry, and friction at the die-workpiece interface. These influence the material flow and grain evolution. For instance, in isothermal forging of titanium alloys – a process conducted at controlled elevated temperature – an insufficiently high blankholder force or suboptimal die design can lead to die underfilling or localized strain concentrations [10], whereas advanced process control (e.g., feedback control of press displacement or force) can mitigate such issues.

Typical applications of forging span aerospace, automotive, and power-generation industries. Forged parts are favored for critical high-stress applications due to their superior internal integrity and mechanical properties. As discussed, open-die forging is used to produce large shafts, rings, and cylinders for turbines, presses, and ship propellers, whereas closed-die forging produces smaller near-net shapes such as automotive crankshafts, connecting rods, and hand tools with minimal post-forging machining [9]. The compressive forming mechanism in forging yields a continuous fiber (grain) flow following the part shape, which enhances fatigue and impact performance. Studies have shown that forging can improve material properties significantly; for example, forging of aluminum matrix composites increases their ultimate tensile strength by consolidating internal porosity, albeit at some reduction in ductility [11]. Modern research in forging focuses on improving die life and process efficiency. One approach to reduce cost in mass production is integrating novel die materials and treatments. For instance, pre-stressed ultra-high-performance concrete (UHPC) dies have been explored for prototype and small-batch forging, as they offer a cheaper, rapid tooling alternative to steel dies [12]. Another active research area is forging process modeling and control – finite element simulations are routinely used to optimize die design and process parameters, predicting material flow, required forces, and potential defect locations [13]. By using numerical optimization and closed-loop control (e.g., adjusting press motion in real time), forging of complex parts can be achieved with fewer trials [14]. Overall, forging's combination of deformation-induced strengthening and achievable near-net shapes makes it a staple for high-performance metal components. Forgings continue to evolve with improved presses, smarter control (including artificial intelligence for defect prediction), and hybrid processes (such as forging combined with subsequent heat treatment) to meet modern demands [13].

Key process considerations also include sufficient heating of the billet is critical in hot forging to ensure the material's flow stress is low enough to fill the die without cracking [15]. However, excessive heat can shorten die life and promote grain growth, so forging temperature is often a compromise. Friction and lubrication at the die interface also strongly affect forging – high friction can cause incomplete filling and localized cooling. Thus, lubricants are used to reduce wear and assist material flow, though they must be chosen to withstand the extreme pressures and temperatures of forging [13]. The strain rate (how fast the material is deformed) influences whether deformation is homogeneous; too high a strain rate can lead to adiabatic heating and localized shear bands, whereas too low a rate may cause excessive heat loss and higher flow stress [15]. Modern forged part production often involves a series of operations: preform forging, intermediate anneals (for very complex shapes), and final forging, followed by trimming of flash (excess material squeezed out in closed-die forging) and possibly heat treatment. Each stage must be optimized to avoid defects. A common forging defect is laps or folds, which occur if the material flows improperly and overlaps on itself; die design modifications and proper material placement can prevent this. Another concern in forging is die wear and failure due to repetitive high loads which require advanced tool steels and hard facing techniques are employed to extend die life [13]. In summary, forging produces parts with excellent strength and reliability by compelling metal to flow into robust shapes, and ongoing research addresses the challenges of tooling durability, energy efficiency, and precision in this ancient yet continually advancing process.

6.2.2 Extrusion

Extrusion is a process in which a billet is forced through a die opening to produce a continuous product with a constant cross-sectional profile [16], such as rods, tubes, or complex aluminum window frames. In direct extrusion (the most common type), a ram pushes the heated billet through a stationary die, whereas in indirect extrusion, the die moves toward a stationary billet, reducing friction. Because the material is under triaxial compression in the deformation zone, extrusion can achieve significant deformation without fracture – the compressive stresses help suppress crack formation, enhancing ductility during processing [17]. Extrusion is widely used for metals like aluminum, copper, and magnesium, particularly to create long shapes with intricate cross sections that would be difficult to form by rolling. For example, aluminum extrusion is used to make automotive chassis components, heat sinks, and architectural frames with hollow sections and stiffening ribs in one piece. The forming mechanism involves shearing and elongation of the billet material as it flows through the die; the extrusion ratio (initial cross-sectional area divided by final area) quantifies the deformation severity. High extrusion ratios require greater forces and can raise the billet temperature due to deformation work and friction. Typically, extrusion of metals is conducted hot (e.g., 450–500 °C for Al alloys, 1,100–1,200 °C for steels) to lower the flow stress and facilitate large deformation. However, even in hot extrusion, the strain rates can be high, so the material may undergo dynamic recrystallization. Indeed, controlling the extrusion speed and temperature is essential: if extrusion speed is too high, excessive heat from plastic work can cause local melting or surface defects; if too low, the material might cool and resist flow, or cause surface sticking to the die [17].

Important process parameters in extrusion include the billet temperature, extrusion speed (ram speed), extrusion ratio, and die geometry (especially the die angle and land length). These parameters determine the required extrusion pressure and the quality of the extrudate. A shallow die angle results in more shear deformation at the die entry but lower redundant work, whereas a very steep die angle can increase the pressure and risk defects like centerline voids (also known as chevron cracking or central burst). Central burst is an internal rupture that can occur when the material in the core of the billet cannot sustain the tensile stress state that develops as it elongates through the die [18]. Studies using finite element analysis have shown that central burst defects are more likely at high extrusion ratios and low friction conditions, and one solution is to use multi-step extrusion (multiple smaller reductions) or a shaped mandrel to support the material and alleviate tensile stresses in the core [18]. Modern extrusion practice incorporates tooling features to minimize such defects; for instance, porthole dies for making hollow extrusions split and re-weld material flow around supports, which must be designed to promote complete bonding without leaving seams or voids. Process innovations such as hydrostatic extrusion – where the billet is encapsulated in a pressurized fluid or soft container during extrusion – have been shown to allow extremely high reduction ratios by applying uniform pressure and reducing friction, thus delaying the onset of internal cracking [18]. This method has been used in research settings to produce ultrafine-grained wires and rods (e.g., hydrostatic extrusion of titanium and steels to refine grain size for improved strength).

6.2.3 Drawing

Drawing, in metalworking, refers to the process of reducing the cross-sectional area of a wire, rod, or tube by pulling it through a converging die [19]. In wire drawing, the material experiences tensile force from the pulling capstan, but deformation occurs under a triaxial stress state as it is squeezed through the die, much like extrusion in reverse. It is one of the oldest and most common cold-forming processes, used to produce wires for electrical conductors, cables, ropes, springs, and numerous other applications [19]. Typically performed at room temperature for most metals (to impart strain hardening for strength), drawing may also be done warm for difficult-to-deform alloys or large cross sections. The drawing die has a profiled opening: an entry angle, a bearing (land) of constant diameter, and often a back-relief exit angle. As the wire is pulled, it undergoes elongation and cross-sectional reduction governed by the area reduction (% reduction per pass) and the die geometry. Key geometric parameters are the die angle, bearing length, and reduction ratio. These factors determine the drawing stress and the quality of the drawn product. For instance, a die with a very short bearing and large angle may lead to non-uniform flow, whereas an overly long bearing increases frictional resistance. An analytical and experimental study by Larsson et al. [19] showed that three die geometry parameters – die angle, bearing length, and reduction (die diameter relative to wire) – all significantly affect the draw force, and classic draw force equations that neglect the bearing length can under-predict the force. By conducting systematic drawing experiments and finite-element simulations, they proposed a new equation incorporating bearing length, which better matched observed forces. This highlights the importance of die design in drawing operations.

During drawing, the material is subject to substantial plastic deformation and thus strain hardens. For example, drawing a steel wire will increase its yield and tensile strength but reduce ductility. Intermediate annealing may be necessary if a large overall reduction is required: a copper wire drawn in many stages from 8 mm rod down to 0.1 mm filament will typically be annealed periodically to restore ductility. Multiple pass drawing is common – wire is drawn through a series of progressively smaller dies, often in tandem on a drawing machine. Modern wire drawing lines can have dozens of dies, with capstans (drums) between them to pull the wire from one die to the next. The synchronization of these capstans (each rotating at a slightly higher speed to account for the elongation) is crucial to maintain proper tension and avoid wire breaks. The drawing speed can be very high for fine wires (tens of meters per second), which poses challenges for lubrication and heat removal. Friction at the die-wire interface converts work to heat; in high-speed drawing this can raise the wire temperature significantly, even though the bulk of the wire remains relatively cool. If lubrication fails or is inadequate, the wire can seize or tear. Thus, lubricants (like stearates for steel wire, or oil-based lubricants for non-ferrous wires) are applied to reduce friction and act as a coolant. However, even with lubrication, the wire and die can experience substantial heating. As a result, the contact conditions (friction coefficient and heat transfer coefficient) strongly influence the temperature distribution in both wire and die during drawing [20]. Excessive temperature rise not only risks lubricant breakdown but can also cause softening that leads to inhomogeneous deformation or surface damage. Hwang and Chang [20] investigated the thermal aspects of wire drawing and found that increases in drawing speed (and thus frictional heating) significantly raise the wire temperature, which in turn can degrade product quality and die life. They noted that higher speed or greater area reduction per pass correlates with greater temperature increase and suggested that controlling the blank (wire) draw-in length and applying cooling could mitigate these thermal effects.

6.2.4 Deep Drawing

Deep drawing is a sheet metal forming process in which a flat sheet blank is radially drawn into a forming die by the action of a punch, creating a hollow three-dimensional shape [21]. In essence, the punch forces the sheet into the die cavity, and the material in the flange region (the outer portion of the blank) must flow inward under a controlled holding force. A blank-holder (or hold-down ring) is used to clamp the sheet blank around the periphery, preventing excessive wrinkling as the material is drawn into the die. In a conventional deep drawing setup, the critical process parameters include the blank holder force (BHF), the die radius and punch radius (which influence the bending/unbending of the sheet), lubrication at tool-sheet interfaces, and the clearance between punch and die (typically a bit more than the sheet thickness to allow smooth flow). BHF must be optimally set as too low BHF allows the flange to wrinkle due to compressive hoop stresses, whereas too high BHF locks the material, causing tearing in the wall or at the punch corner [22]. During drawing, the material in the flange undergoes radial tension and circumferential compression. As the punch pushes in, the flange shrinks in circumference and thickens slightly, while material in the wall of the cup is mostly under tension. The bottom of the cup (attached to the punch face) undergoes stretching. Failure can occur in two primary modes: wrinkling in the flange (a stability failure under compression) or tearing (tensile fracture), often initiating at the punch radius or in the cup wall if the material is over-stretched [22]. A properly chosen BHF and tool geometry aim to prevent both. If wrinkling tendencies are observed, one may increase BHF or employ draw beads (small elevations on the blank holder or die that resist inward flow) to impose additional resistance and tension in the flange. However, increasing resistance to flow makes tearing more likely, so a balance is needed [23]. Modern deep drawing presses can use dynamically adjustable BHF, even varying during the stroke, to optimize conditions: high BHF at the beginning to prevent early wrinkling, then lower BHF later to allow material to flow and avoid tearing. Palmieri et al. [23] developed an in-line control method where the blank draw-in (the movement of the sheet edge) is monitored at key points, and the BHF is adjusted in real time to compensate for friction variability, thereby ensuring a defect-free draw. Such closed-loop control can enlarge the process window for deep drawing, accommodating disturbances like lubrication changes by actively preventing the onset of wrinkling or tearing.

6.2.5 Incremental Sheet Forming

The operating principle of incremental sheet forming (ISF) is simple: a hemispherical or parabolic-shaped punch follows a predefined tool path, gradually deforming the sheet metal blank. This progressive plastic deformation continues until the desired complex geometry of the part is fully formed [24]. Unlike traditional deep drawing,

ISF does not require dedicated dies, enabling low-cost and rapid prototyping of complex geometries [25]. The process generally involves a hemispherical or ball-ended tool that follows a pre-defined tool path, progressively deforming the sheet layer by layer. Variants of the ISF process include single-point incremental forming (SPIF), where a single tool shapes the sheet, and two-point incremental forming (TPIF),, which includes a partial or full die to support the material from below.

ISF offers significant advantages in flexibility and material savings, particularly for customized or small-batch production. However, it typically involves longer cycle times and limited geometric precision compared to conventional methods. Key process parameters include tool path strategy, step depth, feed rate, tool rotation, and wall angle. Excessive wall angles can lead to thinning and fracture, thus determining the formability limits of materials in ISF [26].

Recent studies show that warm ISF improves formability by increasing ductility and delaying necking. For example, Ambrogio et al. [27] demonstrated that heating aluminum sheets during ISF led to better thickness distribution and reduced forming forces. Other research focuses on hybrid processes combining ISF with laser or ultrasonic assistance to enhance material flow and surface finish.

ISF continues to evolve through advancements in robotics, tool design, and simulation techniques, ensuring its relevance in modern manufacturing systems focused on flexibility and customization.

6.3 Hybrid Manufacturing

HM represents a convergence of complementary manufacturing technologies, designed to overcome the limitations of individual processes while exploiting their respective strengths. The growing complexity of engineering components, combined with demands for reduced lead time, improved surface quality, and enhanced mechanical performance, has driven the development of hybrid strategies across both academic research and industrial implementation. As defined by the Lauwers et al. [28], HM processes involve the simultaneous and controlled interaction of process mechanisms and/or energy sources/tools having a significant effect on process performance. The strategic integration of processes such as additive manufacturing (AM), subtractive machining, laser treatment, and thermomechanical operations has enabled the fabrication of intricate, high-performance parts that would be unfeasible using any single process.

HM integrates diverse processes across multiple levels, from energy input to machine configuration and product architecture. Schuh et al. [29] classify it into four main types: (1) simultaneous use of different energy sources in the same zone (e.g., laser-assisted turning); (2) integration of sequential steps into a single operation (e.g., combining extrusion and bending); (3) multifunctional machines combining additive and subtractive methods (e.g., mill-turn centers with direct energy deposition AM); and (4) hybrid products with multi-material or multifunctional structures, such as metal-polymer composites. These categories reflect the flexibility and efficiency of hybrid systems in producing complex, customized, and sustainable components.

The major focus of this chapter is on forming-based HM processes where all typical forming techniques as discussed in Section 6.2 are combined or hybridized with other forming techniques or other manufacturing processes. Hybrid forming techniques based on the assistance from heat and vibration have also been included.

6.4 Forming-Based Hybrid Manufacturing

As shown in Fig. 6.2, forming-based HM combines plastic deformation methods – such as forging, rolling, or stamping – with secondary processes like machining, AM, or surface modification to enhance mechanical properties and dimensional accuracy. The forming stage improves grain structure, strength, and material flow, while follow-up processes refine geometry and surface quality. This hybrid strategy enables the cost-effective production of near-net-shape parts with high structural integrity, reduced defects, and minimal material waste.

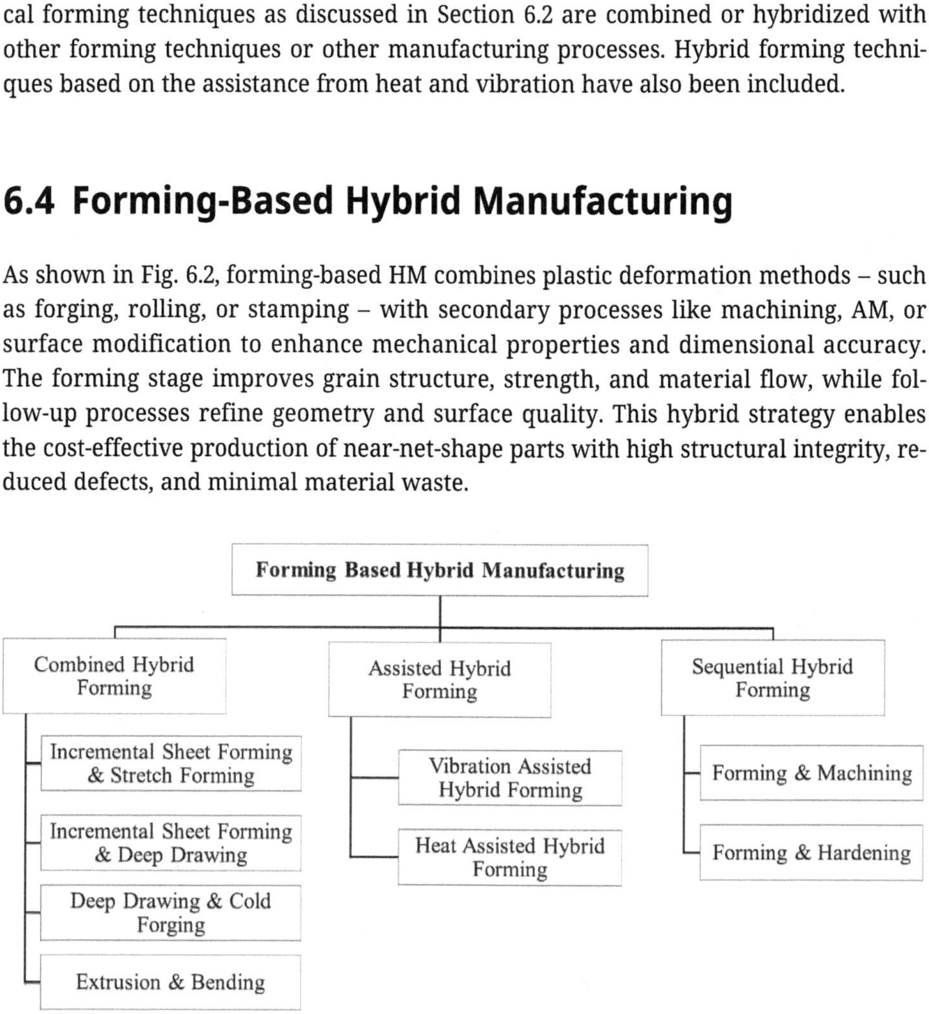

Fig. 6.2: Classification of forming-based hybrid manufacturing.

6.4.1 Combined Hybrid Forming

Combined hybrid forming involves the simultaneous or sequential integration of multiple forming techniques, such as sheet forming and deep drawing or deep drawing and cold forging etc. within a single process chain. This method enhances material flow control, reduces residual stress, and improves microstructural uniformity, enabling the efficient production of complex, high-performance components with optimized mechanical properties.

6.4.1.1 Combined Incremental Sheet Forming and Stretch Forming

The combination of sheet metal forming and stretch forming offers a synergistic approach to improve thickness distribution, geometric accuracy, and overall formability in the manufacturing of complex sheet components [30]. Qin et al. [31] proposed a unified framework for axisymmetric sheet metal forming by modeling it as a combination of ideal stretch forming (Fig. 6.3a). They introduced a dimensionless parameter, X, to quantify stretching relative to drawing. Experimental results show how forming parameters influence shell thickness and guide optimal die design. A hybrid forming method combining stretch forming and asymmetric incremental sheet forming (AISF) was developed to address key limitations of AISF, including excessive thinning, long forming time, and poor geometric accuracy. In their study, Araghi et al. [32] demonstrated that using stretch forming as a preforming stage significantly improves thickness distribution and reduces forming time. Experimental results and finite element simulations confirmed the hybrid process offers superior accuracy and reduced thinning compared to AISF alone.

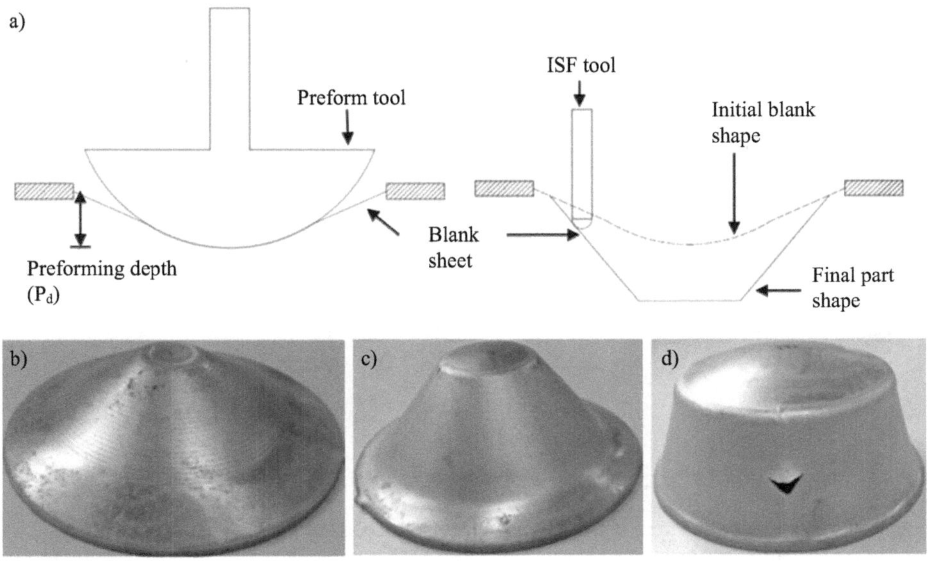

Fig. 6.3: Hybrid AISF with stretch forming: (a) schematic representation and (b) conical frustums with 30°, 50°, and 70° wall angles formed via HISF [30] (with kind permission from Elsevier).

A two-step HISF process combining stretch forming and SPIF was experimentally investigated to address SPIF limitations like localized thinning and poor geometric accuracy. The effects of wall angle, preforming depth, and tool radius on formability and thinning were studied using a full factorial design. Jagtap and Kumar [30] experimentally investigated a two-step hybrid incremental sheet forming (HISF) process that combines stretch forming and SPIF to overcome key limitations of SPIF, such as localized thinning and poor geometric accuracy. Using a simple and low-cost wooden preforming tool, the study analyzed the effects of wall angle, preforming depth, and preform tool radius on minimum thickness and formability through a full factorial design of experiments. The results revealed that the wall angle was the most dominant factor influencing thinning, while appropriate combinations of preforming parameters significantly improved thickness distribution. For instance, at 50° wall angle, a 70 mm preform tool radius and 10 mm preforming depth led to an 18% reduction in thinning compared to SPIF alone (Fig. 6.3b–d).

6.4.1.2 Combined Incremental Sheet Forming and Deep Drawing

Shima et al. [33] introduced incremental deep drawing (IDD) using standard tooling to flexibly form cylindrical cups without shape-specific dies. As shown in Fig. 6.4, a circular punch incrementally displaces the blank, achieving a 1.5 drawing ratio. The key parameters affect formability include punch size, step increment, and holding force, while a counter punch improves drawability and fracture control.

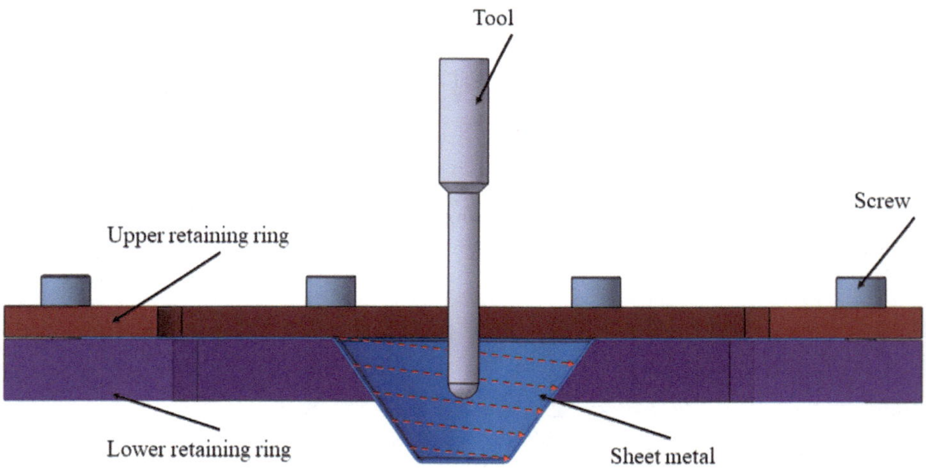

Fig. 6.4: Schematic of working principle of IDD [34]. From *Materials* (MDPI), used under Creative Commons Attribution (CC BY) 3.0 License.

Tandon and Sharma [35] introduced incremental stretch drawing (ISD), a novel hybrid forming method combining ISF with deep drawing to overcome ISF's limitations such as excessive thinning and limited formability at steep wall angles. Their three-stage tool sequence (60, 30, and 13 mm) enable progressive deformation with controlled radial flow. ISD achieved up to 290% thicker walls and supported 90° wall angles without fracture. The final stage uses a small-diameter tool in an outside-in spiral path to refine geometry and accuracy. Experimental results using AL1050 alloy sheets showed improved thickness uniformity, greater depth capability, and better material utilization, making ISD suitable for complex, low-volume, and cost-effective manufacturing. Lim et al. [36] investigated axisymmetrical sheet forming as a simultaneous combination of stretch forming and deep drawing. They introduced a detailed model using a stretch-to-draw-in ratio (*X*-factor) to classify forming behavior and predict cup height. Experimental and theoretical results revealed how blank size, tool profile, and blank-holding force influence formability and optimize cup height.

6.4.1.3 Combined Deep Drawing and Cold Forging

Napierala et al. [37] introduced a novel hybrid metal forming technique termed draw-forging, which simultaneously combines deep drawing and cold forging to manufacture multi-material bulk components. In this process, a deep drawing operation is performed on a steel blank enclosing an aluminum core, which subsequently undergoes cold forging through the same toolset (Fig. 6.5). The process progression involves four stages: (i) initial deep drawing of the blank, (ii) upsetting of the aluminum core and radial expansion of the shell, (iii) redrawing of the shell under compressive loading while the core is forged, and (iv) final forward rod extrusion of the core alone. The experimentally measured force-stroke curve clearly delineates these forming stages, each of which is analytically modeled using elementary plasticity theory. The predicted punch forces show deviations of less than 17% from measured values, validating the mechanical understanding of the process. The joining mechanism between the steel shell and the aluminum core is realized via force- and form-fit, achieving joining strengths exceeding 40% of the shear yield stress of the weaker material.

The bonding mechanisms were further elucidated through cross-sectional analysis and finite element method (FEM) simulations [37]. Surface expansion and contact pressure were identified as key parameters for potential material bonding, though reference process parameters fell short of the thresholds necessary for metallurgical adhesion. Parametric studies revealed the effects of die shoulder angle and extrusion strain on process performance, forming limits, and potential defects such as incomplete head formation, cup fracture, and the formation of interfacial gaps prone to corrosion. A process window was established to delineate the safe and effective operational range. The flexibility of the draw-forging approach was demonstrated through process variants involving chip-based cores, copper-aluminum combinations, and

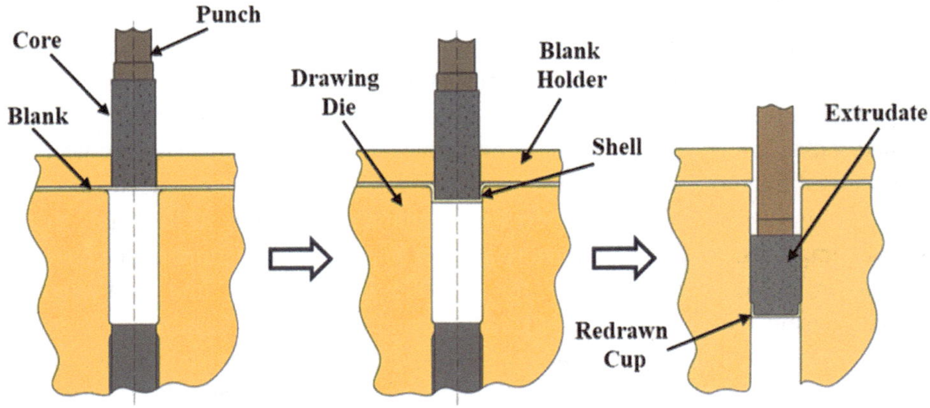

Fig. 6.5: Schematic of draw-forging working principle [37] (modified with kind permission from Elsevier).

multi-step shaft geometries. These results underscore draw-forging as a robust and scalable solution for producing functionally graded multi-material components with tailored surface and structural properties.

6.4.1.4 Combined Extrusion and Bending

Extrusion bending as a sequential HM method integrates direct extrusion with controlled in situ bending to produce curved profiles in a single forming sequence [38]. Unlike traditional multistep approaches involving post-extrusion bending, this hybrid route manipulates metal flow via die design, frictional gradients, or asymmetric extrusion velocities to induce curvature during extrusion itself. This method enhances profile dimensional stability, eliminates the need for intermediate forming dies, and minimizes residual stress concentrations by maintaining continuous material flow under thermomechanically favorable conditions.

Huang et al. [38] implemented a self-bending extrusion method to fabricate L-shaped cross-sectioned AA6061 aluminum alloy rings in a single integrated step. This approach used a specially designed die with an asymmetric exit to induce curvature during extrusion. The process eliminated the need for secondary forming operations by leveraging differential metal flow behavior.

Liu et al. [39] investigated the springback behavior of 2196-T8511 and 2099-T83 aluminum-lithium alloy extrusions during displacement-controlled stretch bending. The authors modeled the bending process over a rigid die and analyzed strain and stress distributions across the cross section. The study examined the influence of process parameters, particularly post-stretch strain, on the residual curvature. Experimental tests and FEM were conducted to evaluate time-dependent springback behavior. It was found that increasing post-stretch strain effectively reduced the radius of springback, thereby improving shape accuracy. Notably, 2196-T8511, with its higher strength, exhibited greater time-dependent elastic recovery compared to 2099-T83. The authors validated their findings by comparing analytical predictions, FEM results, and experimental data, reporting less than 13% error in numerical predictions. This study underscores the importance of post-stretch control in minimizing springback and improving dimensional precision in high-strength Al-Li extrusions.

Clausen et al. [40] also explored the stretch bending of AA7108 aluminum alloy extrusions, focusing on how different tensile loading sequences affect springback and cross-sectional stability. They compared sequences with and without post-stretching. In the first case, (i) the tensile force was either fully applied before bending or (ii) gradually ramped up during the initial bending stage. In the second case, an additional post-stretching force was applied after bending but before unloading. The study found that a constant tensile load equal to the yield strength effectively minimized elastic recovery. However, adding a post-stretching phase further reduced the residual curvature deviation but led to slightly increased local deformation. Springback – defined as the elastic recovery after unloading was quantified using the vertical distance d_0. This parameter served as a key metric in both experimental and FEM analyses to validate the effectiveness of each tensile sequence.

6.4.2 Assisted Hybrid Forming

Assisted hybrid forming combines conventional forming processes with external aids such as heat, vibration, or abrasive to enhance formability, reduce force requirements, and improve material flow. This approach enables precise shaping of complex geometries while minimizing defects and tool wear.

6.4.2.1 Vibration-Assisted Hybrid Forming

Ultrasonic-assisted ISF has shown significant potential in improving formability and surface quality of aluminum alloy sheets. As illustrated in Fig. 6.6, Sun et al. [41] experimentally demonstrated that applying ultrasonic vibration reduces springback and enhances surface finish, with increased amplitude yielding smoother surfaces and more uniform residual stress distribution. Additionally, surface hardness decreased, contributing to better formability and dimensional precision.

Ultrasonic vibration-assisted ISF has been shown to significantly reduce forming forces and improve material formability by leveraging the ultrasonic softening effect. L. Zhang et al. [42] integrated an advanced acoustoelastic-based constitutive model into finite element simulations to analyze the influence of ultrasonic vibration on key process outcomes, including forming force, sheet thickness reduction. The numerical

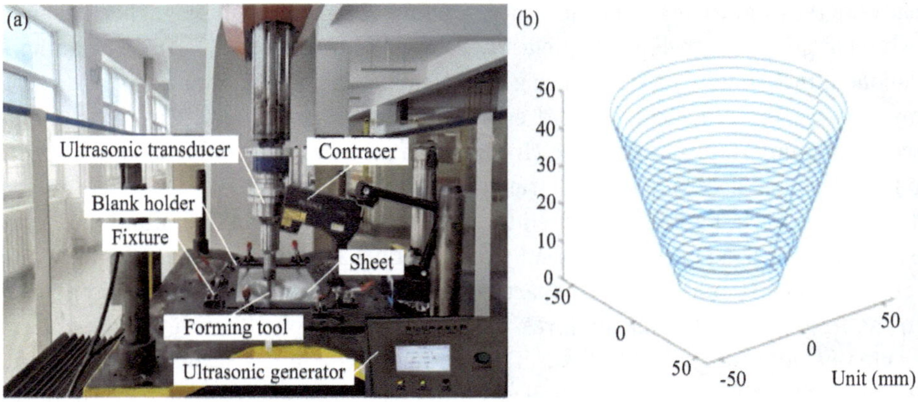

Fig. 6.6: The experimental setup of the ultrasonic-assisted incremental forming process: (a) incremental forming process machine and (b) processing path [41]. From *Symmetry* (MDPI), used under Creative Commons Attribution (CC BY) 3.0 License.

predictions were validated through experimental trials on aluminum 1,050 sheets, demonstrating enhanced geometric accuracy and reduced springback.

Both Su et al. [43] and Alharbi [44] examined the influence of vibration amplitude on the performance of ultrasonic-assisted incremental forming processes, highlighting the critical role of amplitude optimization in improving formability and product quality. Su et al. [43] examined magnesium alloys during ultrasonic vibration-assisted single-point incremental forming (UV-SPIF), analyzing how vibration frequency, amplitude, tool head size, and friction coefficient affect shear stress and thinning rates. They found that a vibration frequency of 20 kHz and a tool radius of 5 mm optimized plastic deformation, though excessive amplitudes (such as 0.16 mm) led to significant thinning and increased fracture risk. Similarly, Alharbi [44] conducted experiments on AA6061-T6 sheets, showing that increasing vibration amplitude (up to 15 µm) reduced forming forces and dimensional deviations, but excessive amplitudes resulted in thinning in critical areas. Alharbi further developed an optimization framework combining ANFIS modeling and PSO algorithms to balance process efficiency with product quality. Both studies emphasize the necessity of controlling vibration amplitude to avoid defects and maximize the benefits of ultrasonic-assisted incremental forming in hard-to-form materials.

6.4.2.2 Heat-Assisted Hybrid Forming

Heat-assisted forming (HAF) is a non-contact method that uses localized thermal energy to improve material formability and enable controlled plastic deformation [45, 46]. In HM, incorporating laser energy into forming processes enhances efficiency and allows precise control by tailoring thermal-mechanical interactions [45]. Min et al. [46] studied laser-assisted robotic roller forming (LRRF) for ultra-high-strength QP1180 steel, addressing its poor room-temperature formability. As shown in Fig. 6.7, the LRRF system features a laser and roller mounted on a robotic arm for synchronized operation. The laser preheats the sheet ahead of the roller, reducing bending forces by up to 43% and springback to under 1°, while achieving an R/t ratio of about 1.2. This thermal softening lowers yield strength near the austenitization temperature, enabling efficient shaping of complex components like electric vehicle battery-pack housings in small-batch production.

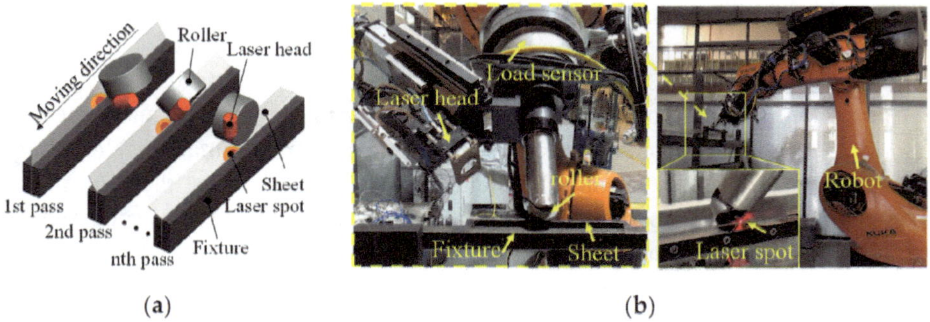

Fig. 6.7: Laser-assisted robotic roller forming: (a) schematic of the forming process and (b) experimental setup [46]. From *Materials* (MDPI), used under Creative Commons CC-BY License.

Guo et al. [47] developed a laser-assisted four-point bending (LAFPB) process specifically for forming high-stiffened aluminum alloy components used in aerospace applications. This hybrid method integrates laser scanning with four-point mechanical bending to enhance localized plastic strain. Their analytical and experimental results showed a 250% reduction in forming forces and springback compared to conventional methods. Additionally, the laser-induced thermal gradient helped redistribute internal stresses, effectively suppressing rib buckling and improving the dimensional accuracy of lightweight, high-strength aerospace structures.

Liu et al. [48] enhanced the LRRF process for ultra-high-strength martensitic steel MS1300 by combining experiments with thermo-mechanical finite element modelling. Their scanning electron microscope analysis confirmed the fully martensitic microstructure of the as-received material. By applying localized laser heating, they reduced forming loads and controlled temperature distribution, resulting in distinct microstructural zones across the bend: newly formed martensite, tempered martensite, and transition regions. This tailored microstructure improved bendability, minimized springback, and reduced fracture risk, demonstrating the effectiveness of LRRF for precision forming.

6.4.3 Sequential Hybrid Forming

6.4.3.1 Forming and Machining

Forming-machining HM integrates plastic deformation techniques with conventional machining operations to enable the production of complex, lightweight, and high-precision components. In this approach, thin features are first created via machining, followed by localized plastic deformation using various forming methods.

In their comprehensive study, Parmar and George [49] systematically explored deformation machining as a novel hybrid approach that merges thin-wall milling and SPIF to fabricate lightweight, high-precision structures. Starting from solid stock, thin geometries were created via conventional computer numerical control (CNC) machining and then deformed using localized forming actions. The authors distinguished two primary deformation modes: bending perpendicular to the tool axis and stretching along it. These modes enable fabrication of structures otherwise unattainable through standalone machining or forming. Feasibility studies confirmed that the deformation forces involved are comparable to machining forces, validating process compatibility with existing CNC machines. Furthermore, the approach offers design flexibility, reduced material usage, and decreased assembly complexity, especially in aerospace and automotive applications. The study also highlights ongoing challenges such as elastic springback, residual stress control, and dimensional inaccuracies, emphasizing the need for further modeling and compensation strategies to enhance precision in hybrid forming-machining systems.

Hybrid forging-machining integrates the strength and grain refinement of forging with the precision and surface quality of machining. While forging improves microstructure, it introduces fatigue-related anisotropies, and machining enhances dimensional accuracy but may cause surface damage. Suárez Fernández et al. [50] examined Ti-6,246 compressor disk components and found that machining introduced beneficial compressive residual stresses, improving fatigue life. However, fatigue variability due to the forged texture persisted. Figure 6.8 presents an example of real component made by forging-machining.

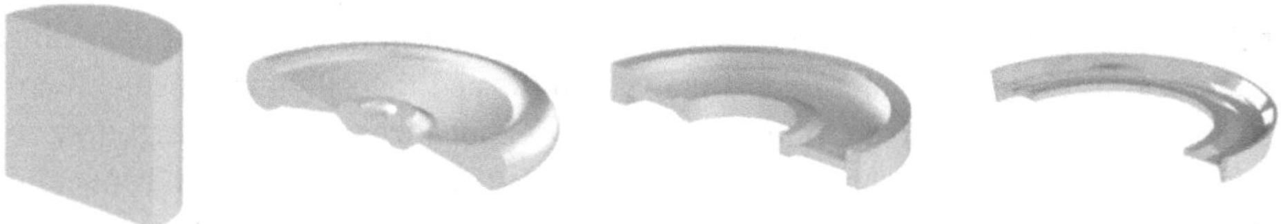

Fig. 6.8: Schematic of the cross-sectional evolution of the component, from left to right: after primary forging, after forging, after roughing to a rectilinear shape, and final machined compressor disk component [50]. From *International Journal of Fatigue* (Elsevier), used under Creative Commons CC-BY License.

Tetsui et al. [51] developed a Ti-42Al-5Mn alloy featuring a stable β-phase to improve hot forgeability and machinability over conventional TiAl. Optimized forging and heat treatments enabled defect-free industrial-scale components. Machinability also improved significantly. Cleanly machined threads and continuous chips were obtained, avoiding the brittle behavior typical of standard TiAl alloys.

6.4.3.2 Forming and Hardening

HM techniques, which integrate forming and hardening heat treatment processes, most notably hot stamping, play a pivotal role in producing high-performance structural components, particularly in the automotive industry [52]. These methods combine plastic deformation and rapid cooling to yield martensitic microstructures directly during forming.

Hot stamping is typically categorized into two main approaches: the direct and indirect methods [53]. In the direct method, a flat sheet blank is first heated in a furnace and then immediately transferred to a water-cooled stamping die, where both forming and quenching occur simultaneously. Conversely, the indirect method involves a preliminary cold-forming step, after which the pre-shaped drawpiece is reheated and subsequently subjected to final forming and quenching in a cooled die. These two techniques differ significantly in terms of process sequence and thermal exposure of the material.

Mori et al. [54] further detailed hot stamping for steels exceeding 1.2 GPa, highlighting its superiority over cold stamping. Their study introduced a new 1,800 MPa-grade steel with finer, uniform martensitic microstructures and improved ductility, enhancing crashworthiness in components like bumper beams. They also reviewed innovations in coatings, heating techniques, and process layouts that improve efficiency and performance.

Rolling hardening is a hybrid surface treatment combining mechanical and thermal effects to induce beneficial microstructural changes, particularly in metastable austenitic steels. It is well-suited for integration into manufacturing lines following soft machining. Meyer et al. [55] investigated cryogenic deep rolling to promote martensitic transformation in X210Cr12 steel. They found that increasing the austenitizing temperature (T) raises the stability of retained austenite, reducing its transformation during loading. They observed enhancement in surface hardening depth and effectiveness through improved thermal-mechanical interactions.

6.5 Summary

Forming-based HM provides a powerful framework to overcome the limitations of conventional forming techniques by integrating complementary processes such as machining, heat assistance, and vibration. These hybrid strategies enhance formability, improve dimensional accuracy, reduce tool wear, and enable the realization of complex geometries that are otherwise unachievable through standalone methods. This chapter has classified these approaches into combined, assisted, and sequential categories, each demonstrating unique benefits in terms of material flow control, energy efficiency, and product quality, while addressing challenges related to springback, thinning, and geometric distortion in advanced material systems. The integration of forming with auxiliary processes opens new possibilities for high-performance manufacturing in demanding sectors like automotive, aerospace, and biomedical engineering. Industrial and academic case studies as discussed in this chapter, demonstrated that hybrid methods not only improve functional performance and surface integrity but also reduce production steps and tooling requirements. As digital control, process modeling, and feedback systems advance, forming-based HM is expected to play an increasingly vital role in achieving precision, customization, and sustainability in metal forming. Future research should focus on real-time control, multi-physics simulation, and scalable implementation to further optimize hybrid process chains for next-generation component manufacturing.

Acronyms and Nomenclature

%	Percent
°C	Degree celsius
µm	Micrometer
3D	Three-dimensional
AI	Artificial intelligence
AISF	Asymmetric incremental sheet forming
Al	Aluminum
AL1050	Aluminum alloy 1050
AM	Additive manufacturing
ANFIS	Adaptive neuro-fuzzy inference system
BHF	Blank holder force
C	Carbon
CNC	Computer numerical control
Cr	Chromium
d_0	Vertical distance parameter used to quantify springback
EV	Electric vehicle
FEA	Finite element analysis
FEM	Finite element method (simulations)
GPa	Gigapascal (unit of pressure)

HAF	Heat-assisted forming
HISF	Hybrid incremental sheet forming
HM	Hybrid manufacturing
IDD	Incremental deep drawing
ISD	Incremental stretch drawing
ISF	Incremental sheet forming
kHz	Kilohertz
LAFPB	Laser-assisted four-point bending
Li	Lithium
LRRF	Laser-assisted robotic roller forming
mm	Millimeter
Mn	Manganese
Mo	Molybdenum
MPa	Megapascal
MS1300	Martensitic steel
PSO	Particle swarm optimization
QP1180	Quenched and partitioned ultra-high-strength steel
R/*t*	Bend radius-to-thickness ratio
SEM	Scanning electron microscopy
Sn	Tin
SPIF	Single-point incremental forming
T	Austenitizing temperature
Ti	Titanium
TiAl	Titanium aluminide
TPIF	Two-point incremental forming
UHPC	Ultra-high-performance concrete
UTS	Ultimate tensile strength
UV-SPIF	Ultrasonic vibration-assisted single-point incremental forming
X	Dimensionless parameter used to quantify stretching relative to drawing
X-factor	Dimensionless ratio
Zr	Zirconium

References

[1] Trzepieciński T. Recent developments and trends in sheet metal forming, Metals. 2020;10:1–53. https://doi.org/10.3390/met10060779.

[2] Folle LF, Lima TN, Santos MPS, Callegari B, Silva BC DOS S, Zamorano LGS. et al. A review on sheet metal forming behavior in high-strength steels and the use of numerical simulations, Metals. 2024;14:https://doi.org/10.3390/met14121428.

[3] Zhu Z, Dhokia VG, Nassehi A, Newman ST. A review of hybrid manufacturing processes – state of the art and future perspectives, Int J Comput Integr Manuf. 2013;26:596–615. https://doi.org/10.1080/0951192X.2012.749530.

[4] Schuh G, Kreysa J, Orilski S. Wie 1+1=3-Effekte in der Produktion Maximiert Werden Können, Z Wirtsch Fabrikbetr. 2009;104:385–391. https://doi.org/10.3139/104.110072.

[5] Sebbe NPV, Fernandes F, Sousa VFC, Silva FJG. Hybrid manufacturing processes used in the production of complex parts: a comprehensive review, Metals. 2022;12. https://doi.org/10.3390/met12111874.

[6] Yue W, Zhang Y, Zheng Z, Lai Y. Hybrid laser additive manufacturing of metals: A review, Coatings. 2024;14. https://doi.org/10.3390/coatings14030315.

[7] Fanani EWA, Surojo E, Prabowo AR, Ariawan D, Akbar HI. Recent development in aluminum matrix composite forging: Effect on the mechanical and physical properties, Procedia Struct Integr. 2021;33:3–10. https://doi.org/10.1016/j.prostr.2021.10.002.

[8] Favi C, Campi F, Mandolini M, Martinelli I, Germani M. Key features and novel trends for developing cost engineering methods for forged components: A systematic literature review, Int J Adv Manuf Technol. 2021. https://doi.org/10.1007/s00170-021-07611-4.

[9] Dindorf R, Wos P. Energy-saving hot open die forging process of heavy steel forgings on an industrial hydraulic forging press, Energies. 2020;13. https://doi.org/10.3390/en13071620.

[10] Ding T, Wei K, Hou Y, Dong X, Huang L, Lee MG. Enhancing the robustness of rib-groove filling and strain homogeneity in the isothermal forging of titanium alloy multi-rib components, Chin J Mech Eng (Engl Ed). 2024;37:https://doi.org/10.1186/s10033-024-01080-y.

[11] Fanani EWA, Surojo E, Prabowo AR, Ariawan D, Akbar HI. Recent development in aluminum matrix composite forging: effect on the mechanical and physical properties, Procedia Struct Integr. 2021;33:3–10. https://doi.org/10.1016/j.prostr.2021.10.002.

[12] Holzer K, Zhang Y, Martinitz L, Hoyer J, Volk W. Experimental study on deep-drawing dies made of pre-stressed UHPC, Materials. 2025;18. https://doi.org/10.3390/ma18020277.

[13] Hawryluk M, Polak S, Rychlik M, Barełkowski A, Jakuć J, Marzec J. Development and implementation of die forging technology eliminating flange welding operations in conveyor driver forging, Materials. 2024;17:https://doi.org/10.3390/ma17133281.

[14] Ding T, Wei K, Hou Y, Dong X, Huang L, Lee MG. Enhancing the robustness of rib-groove filling and strain homogeneity in the isothermal forging of Titanium alloy multi-rib components, Chin J Mech Eng (Engl Ed). 2024;37:https://doi.org/10.1186/s10033-024-01080-y.

[15] Dindorf R, Wos P. Energy-saving hot open die forging process of heavy steel forgings on an industrial hydraulic forging press, Energies. 2020;13. https://doi.org/10.3390/en13071620.

[16] Ferrás AF, De Almeida F, Costa E Silva E, Correia A. Sustainable and optimized production in an aluminum extrusion process, Sustainability (Switzerland). 2025;17:https://doi.org/10.3390/su17094179.

[17] Li M, Yao M, Yang Y, Zhao G, Wang Y, Shu D. et al. Effect of hot extrusion on microstructure, texture, and mechanical properties of Mg-Zn-Mn-0.5Ca alloy, Metals. 2024;14. https://doi.org/10.3390/met14101112.

[18] Lee MG, Korkolis YP. Advances in plastic forming of metals, Metals. 2018;8:https://doi.org/10.3390/met8040272.

[19] Larsson J, Pettersson R, Korin C. Experimental and theoretical examination and development of methods to calculate required pull force in wire drawing, Heliyon. 2024;10. https://doi.org/10.1016/j.heliyon.2024.e39867.

[20] Hwang JK, Chang YC. Effects of contact conditions at wire–die interface on temperature distribution during wire drawing, Processes. 2023;11. https://doi.org/10.3390/pr11020513.

[21] Tricarico L, Palmieri ME. Robust design of deep drawing process through in-line feedback control of the draw-in, Appl Sci (Switzerland). 2023;13. https://doi.org/10.3390/app13031717.

[22] Chen K, Carter AJ, Korkolis YP. Flange wrinkling in deep-drawing: experiments, simulations and a reduced-order model, J Manuf Mater Process. 2022;6. https://doi.org/10.3390/jmmp6040076.

[23] Palmieri ME, Nono Dachille A, Tricarico L. Identification of the optimal blank holder force through in-line measurement of blank draw-in in a deep drawing process, J Manuf Mater Process. 2023;7:https://doi.org/10.3390/jmmp7060190.

[24] Popp GP, Racz SG, Breaz RE, Oleksik VȘ, Popp MO, Morar DE. et al. State of the art in incremental forming: process variants, tooling, industrial applications for complex part manufacturing and sustainability of the process, Materials. 2024;17:https://doi.org/10.3390/ma17235811.

[25] Duflou JR, Habraken AM, Cao J, Malhotra R, Bambach M, Adams D. et al. Single point incremental forming: State-of-the-art and prospects, Int J Mater Form. 2018;11:743–773. https://doi.org/10.1007/s12289-017-1387-y.
[26] Malhotra R, Xue L, Belytschko T, Cao J. Mechanics of fracture in single point incremental forming, J Mater Process Technol. 2012;212:1573–1590. https://doi.org/10.1016/j.jmatprotec.2012.02.021.
[27] Ambrogio G, Gagliardi F, Bruschi S, Filice L. On the high-speed single point incremental forming of Titanium alloys, CIRP Ann Manuf Technol. 2013;62:243–246. https://doi.org/10.1016/j.cirp.2013.03.053.
[28] Lauwers B, Klocke F, Klink A, Tekkaya AE, Neugebauer R, McIntosh D. Hybrid processes in manufacturing, CIRP Ann Manuf Technol. 2014;63:561–583. https://doi.org/10.1016/j.cirp.2014.05.003.
[29] Schuh G, Kreysa J, Orilski S. Roadmap "Hybride Produktion": Wie 1+1=3-Effekte in der Produktion maximiert werden können, Fraunhofer IPT. 2009;104. https://doi.org/10.3139/104.110072.
[30] Jagtap R, Kumar S. An experimental investigation on thinning and formability in hybrid incremental sheet forming process, Procedia Manuf. 2019;30:71–76. https://doi.org/10.1016/j.promfg.2019.02.011.
[31] Qin S, Shang HM, Tay CJ, Mo J. Stretching during axisymmetrical forming of sheet metal, J Mater Process Technol. 1997;63:117–122.
[32] Araghi BT, Manco GL, Bambach M, Hirt G. Investigation into a new hybrid forming process: Incremental sheet forming combined with stretch forming, CIRP Ann Manuf Technol. 2009;58:225–228. https://doi.org/10.1016/j.cirp.2009.03.101.
[33] Shima S, Kotera H, Kamitani K, Bando T. Development of incremental deep drawing process, Metals Mater. 1998;4:404–407.
[34] Popp GP, Racz SG, Breaz RE, Oleksik VŞ, Popp MO, Morar DE. et al. State of the art in incremental forming: process variants, tooling, industrial applications for complex part manufacturing and sustainability of the process, Materials. 2024;17:https://doi.org/10.3390/ma17235811.
[35] Tandon P, Sharma ON. Experimental investigation into a new hybrid-forming process: incremental stretch drawing, Proc Inst Mech Eng B J Eng Manuf. 2018;232:475–486. https://doi.org/10.1177/0954405416645983.
[36] Lim TC, Ramakrishna S, Shang HM. Simultaneous stretch forming and deep drawing in axisymmetrical sheet forming, J Mater Process Technol. 2000; 97(1-3):82–87.
[37] Napierala O, Dahnke C, Tekkaya AE. Simultaneous deep drawing and cold forging of multi-material components: Draw-forging, CIRP Ann. 2019;68:269–272. https://doi.org/10.1016/j.cirp.2019.03.001.
[38] Huang Y, Gao X, Liu Y, Wang Y, Zhao D, Xiao S. et al. Investigation of an L-shaped cross-sectioned AA6061 Aluminum alloy ring via extrusion and self-bending integral processing technology, JOM. 2024. https://doi.org/10.1007/s11837-024-06915-6.
[39] Liu T, Wang Y, Wu J, Xia X, Wang W, Wang S. Springback of extruded 2196-T8511 and 2099-T83 Al-Li alloys in stretch bending, Procedia Eng. 2014;81:981–986. https://doi.org/10.1016/j.proeng.2014.10.128.
[40] Clausen AH, Hopperstad OS, Langseth M. Stretch bending of Aluminium extrusions for car bumpers, J Mater Process Technol. 1999:241–248.
[41] Sun Y, Lu Z, Li C, Wang R, Zhai W. Study on the springback effect and surface property for ultrasonic-assisted incremental sheet forming of Aluminum Alloy, Symmetry. 2021;13(7):1217. https://doi.org/10.3390/sym13071217.
[42] Zhang L, Wu C, Sedaghat H. Ultrasonic vibration-assisted incremental sheet metal forming, Int J Adv Manuf Technol. 2021. https://doi.org/10.1007/s00170-021-07068-5.
[43] Su CJ, Xu TT, Zhang K, Lou SM, Wang Q. Plastic deformation of Magnesium alloy with different forming parameters during ultrasonic vibration-assisted single-point incremental forming, Rare Met. 2022;41:3878–3886. https://doi.org/10.1007/s12598-018-1080-4.

[44] Alharbi N. Experimental study on designing optimal vibration amplitude in ultrasonic-assisted incremental forming of AA6061-T6, Eng Sci Technol Int J. 2022;30. https://doi.org/10.1016/j.jestch.2021.07.004.
[45] Liu Z. Heat-assisted incremental sheet forming: A state-of-the-art review, Int J Adv Manuf Technol. 2018;98:2987–3003. https://doi.org/10.1007/s00170-018-2470-3.
[46] Min J, Wang J, Lian J, Liu Y, Hou Z. Laser-assisted robotic roller forming of ultrahigh-strength steel QP1180 with high precision, Materials. 2023;16(3):1026. https://doi.org/10.3390/ma16031026.
[47] Guo J, Li W, Wan M, Zhao Y, Li C. Design of an innovative laser-assisted four-point bending forming process for high-stiffened structure of Aluminum alloy, Opt Laser Technol. 2021;143. https://doi.org/10.1016/j.optlastec.2021.107313.
[48] Liu Y, Min J, Zhang J, Cai W, Carlson BE, Bobel AC. et al. Laser-assisted robotic roller forming of an ultrahigh strength martensitic steel, J Manuf Process. 2022;82:192–202. https://doi.org/10.1016/j.jmapro.2022.07.066.
[49] Parmar PLL, George PMM. Recent advances and research in new hybrid technique deformation machining, 2024;44.
[50] Suárez Fernández D, Wynne BP, Crawforth P, Fox K, Jackson M. The effect of forging texture and machining parameters on the fatigue performance of Titanium Alloy disc components, Int J Fatigue. 2021;142. https://doi.org/10.1016/j.ijfatigue.2020.105949.
[51] Tetsui T, Shindo K, Kaji S, Kobayashi S, Takeyama M. Fabrication of TiAl components by means of hot forging and machining, Intermetallics. 2005;13:971–978. https://doi.org/10.1016/j.intermet.2004.12.012.
[52] Merklein M, Svec T Hot stamping: Manufacturing functional optimized components, Prod Eng. 2013;7:141–151. https://doi.org/10.1007/s11740-012-0429-z.
[53] Wróbel I, Skowronek A, Grajcar A A review on hot stamping of advanced high-strength steels: Technological–metallurgical aspects and numerical simulation, Symmetry. 2022;14. https://doi.org/10.3390/sym14050969.
[54] Mori K, Bariani PF, Behrens BA, Brosius A, Bruschi S, Maeno T, et al. Hot stamping of ultra-high strength steel parts, CIRP Ann Manuf Technol. 2017;66:755–777. https://doi.org/10.1016/j.cirp.2017.05.007.
[55] Meyer D, Brinksmeier E, Hoffmann F Surface hardening by Cryogenic Deep Rolling, Procedia Eng. 2011;19:258–263. https://doi.org/10.1016/j.proeng.2011.11.109.

Vivek Kumar Tiwary, Arunkumar P., Vinayak R. Malik

7 Hybrid Additive-Subtractive Manufacturing: The Integration of 3D Printing and Precision Machining for Advanced Manufacturing

Abstract: Hybrid additive-subtractive manufacturing (HASM), also referred to as large-scale additive manufacturing, fused granulate fabrication, or big area additive manufacturing, represents a transformative approach in modern manufacturing, combining the capabilities of additive manufacturing (AM), such as 3D printing, with the precision and surface quality achievable through subtractive machining (SM) processes like computer numerical control machining. This integrated technique allows for the production of complex geometries with high accuracy and enhanced surface finishes in a streamlined workflow. By leveraging the strengths of both AM and SM, hybrid manufacturing minimizes material waste, reduces production time, and enables the creation of multi-material components. It is especially valuable in industries such as tool and die, aerospace, automotive, and medical devices, where performance and precision are paramount. This chapter explores the principles, processes, advantages, and applications of HASM. It includes insights from an industrial perspective, along with recent technological innovations and advancements in HASM. Furthermore, several case studies are presented, highlighting critical challenges and the future scope of HASM, and demonstrating its role in advancing industrial fabrication and driving innovation in part design and production.

Keywords: CAD, CNC machine, hybrid manufacturing, 3D printing, prototype, surface quality

7.1 Introduction

Large format additive manufacturing (LFAM) is a form of 3D printing technology specifically designed for creating large-scale objects or structures. Machines or equipment that can print volumes greater than 1 m^3 are typically referred to as LFAM [1]. Traditional 3D printing uses materials like plastic, metal, or composites to build objects layer by layer, whereas LFAM uses polymer pellets [2]. One of the critical limitations of desktop 3D printers, the bed size, can also be circumvented by this technique.

The pellet-based additive manufacturing (AM) process eliminates the need for a prior filament-extrusion step, potentially reducing costs by up to tenfold and in-

 | https://doi.org/10.1515/9783111704456-007

creasing production speed by as much as 200 times. Moreover, polymer filaments offer limited material options, whereas pellets are available in nearly all industrial polymers, making them a more versatile and attractive choice [3].

LFAM represents a progression in AM technology that has made possible the development of bigger and more complex products, as shown in Fig. 7.1. This approach could revolutionize high-volume production industries with precise engineering and careful material choices. LFAM technology faces challenges in balancing surface quality with printing speed. Additionally, LFAM printers are expensive, and encounter issues related to scalability, reliability, and process repeatability [4].

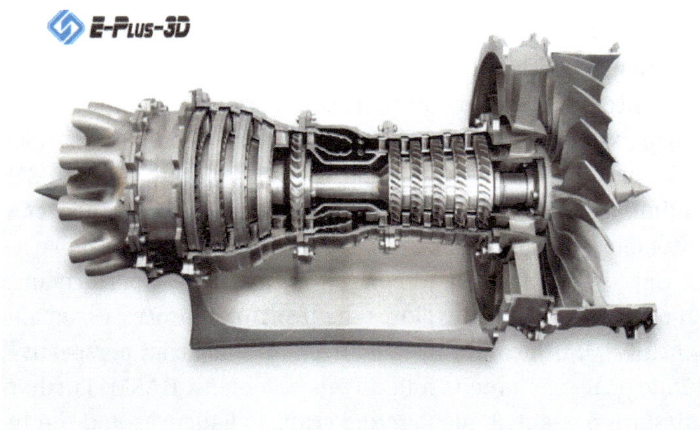

Fig. 7.1: LFAM for aerospace parts [5] (courtesy of EPlus3D).

Hybrid additive-subtractive manufacturing (HASM) is a recently developed "near net shape" methodology of product development and manufacturing that begins with high-speed printing followed by cutting/trimming to the final shape and size. This method is the fastest and most efficient approach to building large structures with 3D printing technology. Figure 7.2 demonstrates the fabrication of a part by hybridization of 3D printing to give an overall form and CNC machining to finally shape [6]. The huge bulk of thermoplastic composite materials, including high temperature materials suitable for molds and tooling can be processed into parts by this technique.

An HASM is a novel idea that differs from a normal desktop 3D printer in three ways. First, it has a facility for pellet inputs, second, the printing area is large (larger than 1 cubic meter), and third, both additive and subtractive operations are performed simultaneously. Figure 7.3 also illustrated the final part shaping by the combination of AM and computer numerical control (CNC) machining.

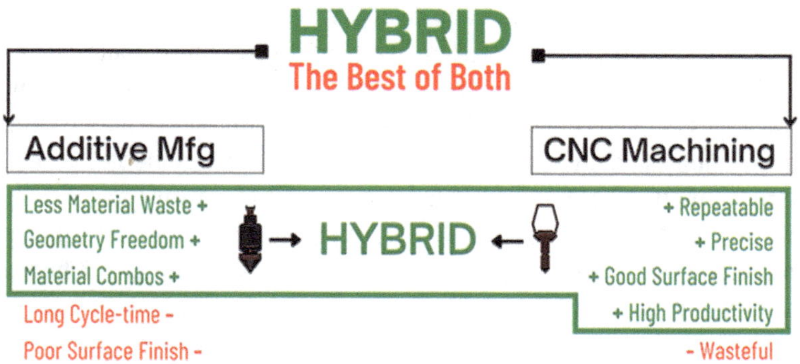

Fig. 7.2: The process of hybrid additive-subtractive manufacturing [7] (courtesy of Hybrid Manufacturing Technologies).

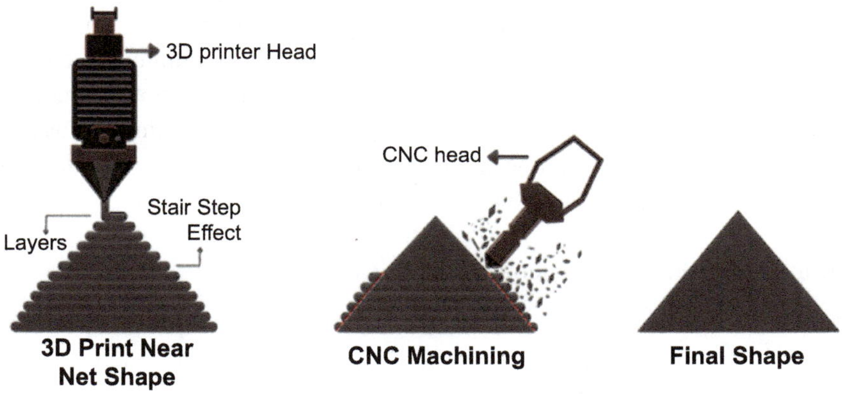

Fig. 7.3: The hybrid approach in additive manufacturing.

7.2 Principle of Hybrid Additive-Subtractive Manufacturing

AM plays a central role in driving the transformation from the fourth to the fifth industrial revolution. The technology involves creating parts by sequentially adding layers of material, enabling the production of highly complex geometries using various materials such as metals, polymers, composites, and ceramics. Although design freedom, less material wastage, low energy consumption, shorter production times, and shorter supply chains are key advantages, issues related to the poor surface finish, anisotropy, and low mechanical strength still need to be addressed.

Subtractive machining (SM) removes material from a solid workpiece to create parts with desired shapes and dimensions. Using tools like mills, lathes, and grinders, this process is highly precise and versatile, capable of working with metals, plastics, ceramics, etc. SM offers benefits like high precision, excellent surface finish, material versatility, and scalability while material wastages, tool wear, high energy usage, and setup complexity are the issues of the technology [8]. Hybrid additive-subtractive manufacturing (HASM), which combines additive and subtractive manufacturing, emerges as a solution to address the limitations inherent in each individual manufacturing method.

The principle of HASM, as illustrated in Fig. 7.4, integrates additive and subtractive processes to capitalize on their strengths and overcome their limitations. It begins with creating a detailed computer-aided design (CAD) model and planning the manufacturing sequence, balancing material addition, and precision removal. The process starts with AM, where material is deposited layer by layer to form the basic geometry, including complex shapes and internal structures. This is followed by SM, which removes excess material to refine dimensions, improve surface finishes, and achieve tight tolerances. The process can alternate between additive and subtractive phases, allowing iterative refinement for specific regions or features. Advanced systems may include intermediate inspections using sensors or 3D scanning to ensure accuracy throughout the cycle. Once the desired geometry and finish are achieved, the part undergoes final inspection and post-processing, such as polishing or thermal treatment, to meet functional and aesthetic requirements. This approach delivers high-quality, geometrically complex components efficiently and accurately.

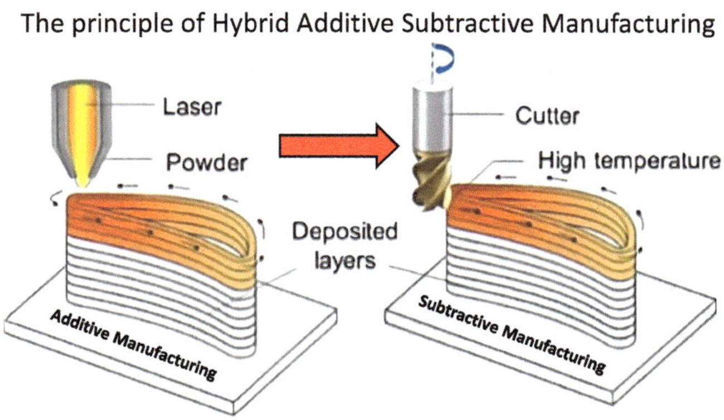

Fig. 7.4: The principle of HASM process [9]. From the *International Journal of Advanced Manufacturing Technology* (Springer), used under Creative Commons Attribution (CC BY) 4.0 License.

7.3 The Process of Hybrid Additive-Subtractive Manufacturing

The process of HASM is divided into three key steps: software preparation, printing, and trimming. Each step is detailed below:

(a) **CAD modeling of the desired part**
The process begins with a preparing the CAD model of the desired part, created in an industry-standard solid, surface, or mesh file format. This model is imported into specialized print software, which generates both the printing and trimming programs. These programs guide the additive and subtractive phases, ensuring precision and seamless integration.

(b) **Slicing**
This is a process of converting a digital design into a set of instructions that a robotic system can follow to manufacture a part. During this step, the 3D model of the design is broken down into thin horizontal layers, and the resulting data is translated into machine-readable code. This code specifies the precise movements of the robotic tool head, including its path, orientation, and speed, as well as the extrusion rates required to deposit material during AM. The slicing process also accounts for parameters like layer thickness, material flow, and adjustments for specific features such as overhangs or supports. By carefully optimizing these factors, slicing ensures that the design can be accurately and efficiently realized while minimizing errors and material waste.

(c) **Printing**
In this step, thermoplastic material is heated till it becomes soft and pliable, then extruded as a continuous bead. Layer by layer, the material is deposited to build the part, with each new layer fusing seamlessly with the previous ones to create a solid, strong, and void-free structure. This additive phase forms the basic geometry and complex features of the part.

(d) **Annealing**
After the mold is printed, it is heated to a high temperature to release residual stresses caused by the AM process. These stresses arise from rapid cooling and solidification during printing, which can affect the mold's accuracy and strength. The heating process, known as stress-relief annealing, helps the material relax and ensures the mold maintains its structural integrity. This step is essential to prepare the mold for the further use, without any deformation or failure.

(e) **Trimming**
Trimming is performed employing a three-axis CNC router mounted on a discrete gantry that shares the same overhead rails as the printing gantry. The router, typically equipped with a 6 kW spindle capable of 3,000–24,000 rpm, ensures precise material removal. This phase refines dimensions, enhances surface finishes, and achieves tight tolerances, completing the manufacturing process.

(f) **Surface preparation**
The next step includes sealing the surfaces which is an important post-processing step to improve part quality. During this process, any existing voids or imperfections on the surface are filled, ensuring a smooth and uniform finish. This enhances the part's structural integrity, durability, and resistance to environmental factors like moisture or corrosion. Sealing also prepares the surface for further

treatments, such as painting or bonding, ensuring that HASM produced components meet the performance standards required in industries like aerospace or marine.

A case of HASM with the details of complete procedure is depicted in Fig. 7.5.

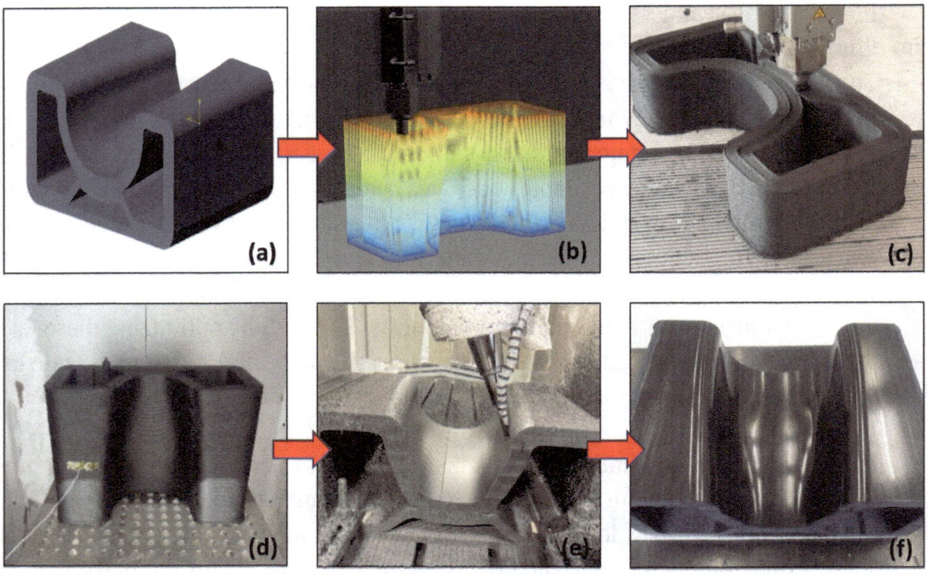

Fig. 7.5: The process of hybrid additive-subtractive manufacturing. (a) CAD modeling, (b) slicing, (c) printing, (d) annealing, (e) trimming, and (f) surface preparation [10] (courtesy of Royal Netherlands Aerospace Centre).

7.4 Applications of Hybrid Additive-Subtractive Manufacturing

HASM is being employed in various sectors ranging from tool and die making to aerospace, automotive, and biomedical applications. The following section attempts to highlight the applications of HASM in the various fields.

7.4.1 Aerospace: Lightweight Structures and Complex Geometries

Innovative manufacturing operations are of paramount importance in the aerospace sector these days to develop fuel-efficient and environmentally sustainable aircraft due to the increased air travel demand and progress in space exploration [11]. The creation and fabrication of aerospace molds, including composite cure tooling, is essential for guaranteeing the precision and dependability of aircraft components. Furthermore, as the demand for sophisticated, lightweight, and geometrically complex aerospace components increases, the industry necessitates tooling, manufacturing equipment, and procedures that can adapt to these requirements [12]. In the past, aerospace tooling has depended on traditional materials and production processes, such as CNC metal machining and fiber-reinforced composite layup, which can involve expensive and time-consuming procedures, limiting the intricacy of designs. Now, with the advent of HASM, rapid production of aerospace parts, tooling, and components has been revolutionized. One such example is depicted in Fig. 7.6.

Fig. 7.6: A critical aerospace engine cowlings component being fabricated by HASM [13] (courtesy of CEAD B.V., Delft, The Netherlands).

7.4.2 Naval and Marine Architecture Applications

HASM is transforming the way by which the design and building of marine vessels happens. By combining the strengths of 3D printing, which excels at creating complex shapes and minimizing waste, with traditional machining for fine-tuning and precision, HASM opens up new possibilities for naval architecture. This technology is ideal for producing parts such as propellers, hull reinforcements, and other critical components that need to be both strong and lightweight [14]. It also makes it easier to test and refine designs quickly, allowing for faster innovation and customization. As shipbuilders face growing demands for efficiency, sustainability, and advanced performance, HASM is proving to be a powerful tool that bridges creativity and practicality in modern shipbuilding. One such example of a boat hull being manufactured by HASM technique is depicted in Fig. 7.7.

The image effectively illustrates the integration of additive and subtractive methods, showcasing how these processes complement each other to produce complex, large-scale components like a boat hull with precision and efficiency. This setup highlights the potential for advanced manufacturing in the marine industry, enabling rapid prototyping, and reduced material waste.

Fig. 7.7: A boat hull being manufactured by the HASM technique [15] (courtesy of Thermwood).

7.4.3 Tool and Die Industry

HASM is revolutionizing the tool and die industry by combining the strengths of 3D printing AM and CNC machining (subtractive manufacturing). This hybrid approach enables rapid prototyping, production of complex geometries, and on-demand manufacturing, significantly reducing lead times and costs. Additive processes create near-net-shape parts with minimal material waste, while subtractive methods ensure precision and high-quality surface finishes. HASM is particularly valuable for repairing and maintaining tools and dies, extending their lifespan by adding and refining material as needed [16]. Key advantages include enhanced design flexibility, material efficiency, reduced costs, improved tool performance, and the ability to produce advanced internal features like conformal cooling channels. By combining speed, precision, and versatility, HASM provides a transformative solution for the precision-driven demands of the tool and die industry. As

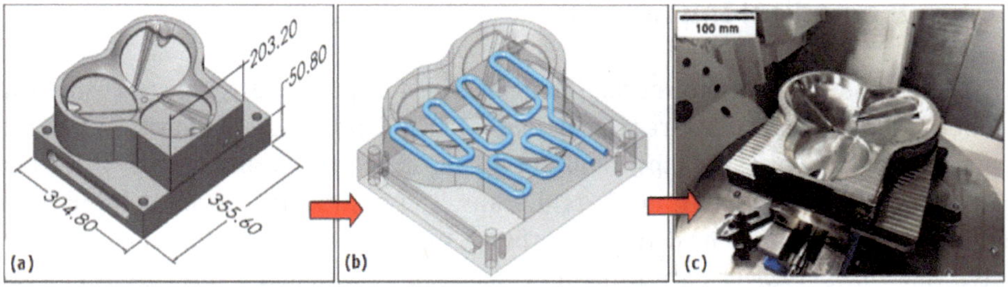

Fig. 7.8: HASM (DED process) being employed to fabricate a mold for conformal cooling [17]. From *Journal of Manufacturing and Materials Processing* (MDPI), used under Creative Commons Attribution (CC BY) 3.0 License.

shown in the figure, mold for conformal cooling channels being fabricated via HASM is depicted. Figure 7.8a shows the CAD model, whereas Fig. 7.8b shows the conformal cooling channels and Fig. 7.8c shows the finished mold after HASM.

7.4.4: Automotive Prototyping and Custom Components

HASM is highly effective in the automotive industry for prototyping and producing custom components. By combining AM (3D printing) with traditional subtractive techniques like CNC machining, HASM allows for rapid creation of complex, highly precise prototypes. This enables designers to test and refine parts more quickly during the development phase. Furthermore, it facilitates the fabrication of bespoke parts featuring intricate geometries that conventional techniques struggle to produce. HASM ensures high-quality finishes and mechanical properties, making it an ideal solution for both experimental prototypes and tailored automotive parts, as shown in Fig. 7.9.

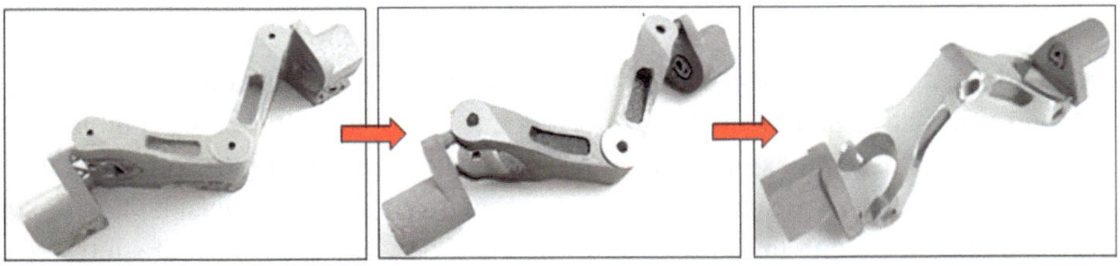

Fig. 7.9: HASM (EBM process) being employed to fabricate a critical automotive component [18] (with kind permission from Taylor & Francis).

7.4.5: Biomedical Implants and Patient-Specific Devices

HASM can be highly useful in the biomedical sector, particularly where customization, speed, and scalability are crucial. As shown in Fig. 7.10, LFAM allows for patient-specific implants (e.g., for hips, skulls, or spinal structures) to be created with exact anatomical accuracy based on CT/MRI scans. These implants can be tailored to fit unique bone structures or defects, improving surgical outcomes and patient comfort [19]. Nowadays, surgeons are using life-sized anatomical models produced by LFAM to visualize complex cases before performing surgery. This helps them in preparing surgical strategies, reducing time in the operating room and increasing surgical precision. LFAM enables faster and more cost-effective production of custom prosthetics and orthotic devices, especially for larger body parts like legs or torso braces [20]. However, the surface finish, biocompatibility, and regulatory hurdles for implantable devices are the key challenges LFAM must address.

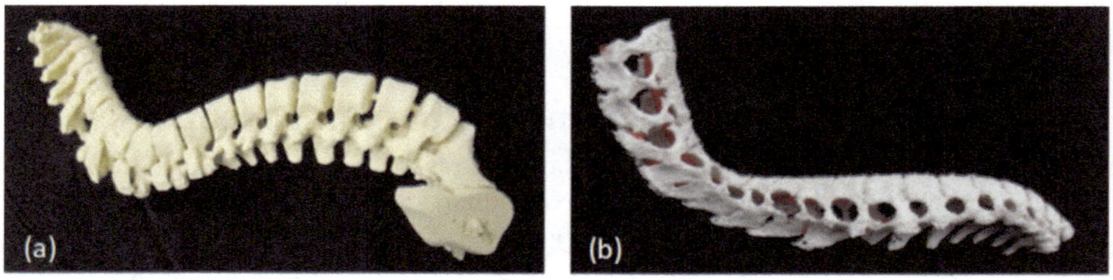

Fig. 7.10: Application of LFAM to make a complete spinal cord for the patient [21]. From *Journal of Medical Engineering* (Wiley), used under Creative Commons Attribution (CC BY) 4.0 License.

7.5 Industrial Perspective on Hybrid Additive-Subtractive Manufacturing

The HASM market is experiencing significant growth, driven by the integration of additive and subtractive manufacturing processes into single machines. This combination offers enhanced design flexibility, rapid prototyping, and efficient production capabilities. Sectors such as aerospace, defense, and automotive are leading the adoption of hybrid manufacturing due to the seamless transition between additive and subtractive processes, which streamlines production and reduces time-to-market [22].

Hybrid manufacturing processes contribute to a reduced carbon footprint by minimizing waste and emitting fewer harmful pollutants compared to traditional manufacturing methods. The market is characterized by a variety of advanced technologies and features, leading to a fragmented landscape. The incorporation of multi-axis systems in hybrid machines addresses limitations of standalone AM, enabling complex part production with improved efficiency and material savings. Some of the leading companies in the market that are forefront in developing and implementing hybrid AM technologies, contributing to the market's expansion are depicted in Tab. 7.1.

DMG Mori offers four hybrid machines – Lasertec 65, 125, 3000, and 6600 – using DED with a powder nozzle [24]. These machines combine AM and milling for high-quality, finished 3D parts in industrial applications.

The Mazak INTEGREX i-400 AM is a Japanese hybrid 3D printer using laser beam DED for steel parts [28]. It features full five-axis machining, −30°/+210° B-axis movement, C-axis contouring, and a programmable NC tailstock for versatile part production.

Cead B.V. provides turnkey hybrid AM systems combining fused granulate fabrication (FGF) with CNC milling and trimming [13]. Their Flexbot and FlexCube platforms offer both additive and subtractive capabilities, integrated with 3D scanning, process logging, and parameter data management. Flexbot includes a 14-slot tool bay with a scanning head; FlexCube features a 10-slot bay. Extruder options range from 12 to 84 kg/h (E25, S25, E40, and E50), and all systems use Hiteco milling spindles. With over 170 global installations, their largest system, is in Abu Dhabi, spanning 40 m × 4 m.

Thermwood provides one of the most diverse ranges of large-scale AM systems currently in the market [23]. Their large-scale additive manufacturing (LSAM) series is designed to fabricate medium to extremely large components using reinforced thermoplastic composites. Additionally, their cut layer additive system creates similarly sized parts from a broad selection of materials that are not printable but can be rapidly machined. Both technologies are capable of producing industrial tools, master models, molds, and production fixtures across various sectors, including aerospace, automotive, marine, and foundry industries. Table 7.1 depicts various commercially available machine tools featuring hybrid manufacturing and their capabilities.

7.6 Technological Innovations and Advances

The pace of technological advancements in large-format 3D printing is so rapid that there exists a need for a comprehensive review that encapsulates the recent and significant developments in this field [4].

An emerging but less widely recognized approach in HASM involves the fabrication of sandwich structures. This method entails depositing cores with varying densities, as described in a patent by [31]. This approach enables the production of lighter structures and the customization of mechanical properties based on the anticipated load in each specific area.

A key advancement in HASM is AI-driven process monitoring [32]. Artifical intelligence (AI) analyses sensor and camera data in real time to detect issues like warping or extrusion defects, automatically adjusting print parameters to maintain quality. Machine learning (ML) also enables predictive maintenance, reducing downtime and

Tab. 7.1: Commercially available machine tools featuring hybrid manufacturing capabilities.

S. no.	Company	Product	Additive process	Subtractive process	Reconfiguration method	Printing software	References
1	Deltasys E Forming	DE LFAM 1020	Fused filament fabrication	Five-axis CNC router	Automatic tool changing (ATC)	Siemens NX or Fusion 360	[6]
2	Thermwood Corporation	Large format additive manufacturing (LFAM)	Fused filament fabrication	Five-axis CNC router	3,000–24,000 RPM ATC Spindle	LSAM Print 3D	[23]
3	DMG Mori	Lasertec 65 3D Hybrid	LED process	Full five-axis milling machine	Automatic shuttle system	SIEMENS NX Hybrid CAM	[24]
4	Ingersoll Machine Tools	MasterPrint® 3X	Fused filament fabrication	Five-axis milling head	18,000 rpm max spin	Ingersoll proprietary software	[25]
5	Phillips	Phillips Additive Hybrid	Wire laser directed energy deposition (DED)	Five-axis CNC router	Automated reconfiguration method	Fusion 360/Siemens NX	[26]
6	Matsuura Corporation	Lumex Avance – 25, Matsuura, Machinery Corp.	Powder bed fusion (PBF) method	Three-axis CNC machining	Automated	LUMEX CAM Software	[27]
7	Cead B.V.	Flexbot and FlexCube	Fused granulate fabrication (FGF) technology	CNC milling and trimming	Hiteco	Modular software approach	[13]
8	Mazak Corporation	INTEGREX i-400 AM	Wire arc additive manufacturing (WAAM)	Multi-axis CNC milling and turning	Automatic tool changing (ATC)	Mazak SmoothX Control	[28]

9	Optomec	LENS Hybrid Series	Laser engineered net shaping (LENS)	Multi-axis CNC machining	Automatic nozzle retraction before CNC machining	Siemens NX	[29]
10	Diversified Machine Systems (DMS)	Hybrid CNC 5-Axis	Fused filament fabrication	Five-axis CNC milling	Automatic print head retraction	Mastercam Hybrid Manufacturing Module	[30]

waste. By enhancing precision and efficiency, AI is making LFAM more reliable and practical for large-scale industrial applications. Another advancement in HASM is related to involve digital twins, which serve as real-time virtual replicas of the physical manufacturing process [33]. This integration enables real-time monitoring, simulation, optimization, and control of both additive and subtractive phases. Key benefits include real-time feedback and control, predictive maintenance and defect detection, optimized process planning, and improved product quality with enhanced dimensional accuracy [34]. However, challenges such as the high computational load required for real-time simulation must be addressed for this integration to be fully realized. HSAM is rapidly advancing with the integration of ML, which significantly enhances process intelligence, adaptability, and overall efficiency. ML algorithms, trained on large volumes of process data, are being applied across various stages of HASM to optimize operations, minimize errors, and improve product quality [35]. These ML-driven innovations are especially impactful in high-value and complex manufacturing sectors such as aerospace, biomedical, and automotive, where reliability and customization are essential. Nevertheless, challenges persist, including data standardization, model generalizability, and the need for real-time computational capabilities. HSAM and Internet of Things (IoT) can be yet another area of collaboration which can enable smart, connected manufacturing systems capable of real-time data acquisition, analysis, and control [36]. In this context, IoT involves embedding sensors, actuators, and communication modules within HASM machines to collect critical process data – such as temperature, vibration, tool wear, surface roughness, melt pool characteristics, and machine health parameters during both additive and subtractive operations [37].

HASM presents a promising approach to addressing sustainability challenges in advanced manufacturing. By integrating additive and subtractive processes within a single system, HASM enables near-net-shape production, significantly reducing material waste and machining time. The capability to repair or modify existing components extends product life cycles and decreases the demand for raw materials. Additionally, process consolidation minimizes energy consumption, transportation, and handling, contributing to lower overall environmental impact [35, 36]. The use of recycled or sustainable feedstocks further enhances the ecological benefits of HASM, positioning it as a key enabler of resource-efficient, low-carbon manufacturing aligned with circular economy principles.

7.7 Case Studies and Real-World Implementations

7.7.1 HASM-Powered Fully 3D-Printed Monocoque for CO_2 Neutral Car – ZEM

The students of TU/ecomotive team from Eindhoven University of Technology developed a ground-breaking concept car, ZEM (zero emission mobility), aiming for CO_2 neutrality across all its development phases [13]. Leveraging HASM and sustainable materials, the team addressed the critical challenges posed by climate change and demonstrated the potential for innovation in the automotive industry.

Europe aims to cut CO_2 emissions by 55% by 2030 and reach climate neutrality by 2050 as part of its ambitious climate objectives [13]. For the automotive industry, this necessitates drastic changes in materials, manufacturing, and overall vehicle life cycle processes. TU/ecomotive embraced these challenges, seeking to create a vehicle that exemplifies sustainability throughout its design, production, use, and disposal phases. Material selection, manufacturing constraints and life cycle considerations were the key challenges identified.

To achieve CO_2 neutral production, the team utilized E25 extruder HASM for ZEM's monocoque (vehicle structure in which the chassis is integral with the body) and body panels employing S25 extruder. Recycled PETG was first chosen for its sustainability and versatility, made from recycled cooking oils. After extensive FEM analysis, the team selected Mitsubishi Chemical Europe's Carbon-P, a recycled PETG reinforced with carbon fibers, for its strength and circularity. The monocoque design was developed with HASM-3D printing in mind, avoiding excessive overhangs to minimize production failures. The structure was modeled symmetrically along the *ZY* plane to simplify and streamline production, as shown in Fig. 7.11.

HASM method allowed precise fabrication with minimal waste, incorporating circular plastics that could be reused in future projects. The project culminated in a fully 3D-printed monocoque that met the team's sustainability and performance goals wherein circular plastics and AM drastically reduced waste and embodied carbon. AM enabled the creation of intricate designs previously constrained by traditional methods and the successful production of ZEM validated HASM-3D printing as a viable method for automotive manufacturing. The team's efforts proved that sustainable mobility is achievable, providing a blueprint for the automotive industry to reduce its environmental footprint [13].

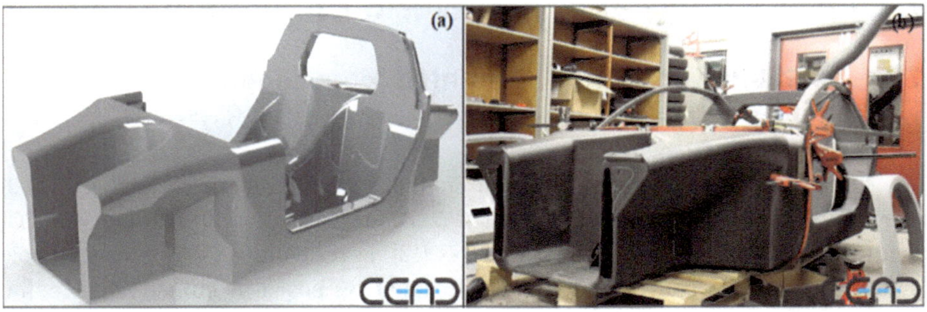

Fig. 7.11: (a) The initial CAD file and (b) the monocoque fabricated using HASM technique [13] (courtesy of CEAD B.V., Delft – The Netherlands).

7.7.2 HASM in Shipbuilding: Northern Light Composites' Sustainable Hull Innovation

Northern Light Composites (NLComp), a pioneer in sustainable shipbuilding, collaborated with Breton to address a formidable challenge: the creation of molds for the prototype of EcoFoiler, the world's first recyclable foiling hull [38]. This ambitious project aimed to design an innovative and eco-friendly hull using advanced thermoplastic materials, without compromising on precision or quality. The partnership exemplifies how cutting-edge technology and sustainable materials can merge to revolutionize traditional industries.

NLComp's objective was to develop a recyclable hull that would set a new benchmark in sustainable marine engineering. However, the intricate design and precision requirements of the EcoFoiler prototype posed significant technical hurdles. Addressing these challenges necessitated the integration of Breton's expertise in advanced manufacturing with NLComp's pioneering thermoplastic materials.

Breton deployed its state-of-the-art large-scale AM system, the Breton Genesi, to meet the demands of the project. The Breton Genesi is equipped with the capability to 3D print on work surfaces measuring approximately 4,000 x 10,000 mm and up to a height of 2,000 mm. Leveraging advanced ML algorithms and AI, the system optimized every stage of production to ensure exceptional mechanical and aesthetic properties.

The collaboration yielded large, high-precision components for the EcoFoiler hull while minimizing production costs and environmental impact (Fig. 7.12). The innovative approach not only demonstrated the potential of recyclable thermoplastic materials in shipbuilding but also highlighted the efficiency and sustainability of large-scale AM.

This case study underscores the transformative potential of combining sustainable materials with advanced manufacturing technologies. The success of the EcoFoiler project by NLComp and Breton illustrates how innovation can drive sustainability in traditionally resource-intensive industries like shipbuilding [38].

Fig. 7.12: (a) Post-additive manufacturing, (b) partially milled, and (c) fully finished product [38] (courtesy of Breton SpA – Italy).

7.8 Critical Challenges in HASM Applications

HASM, which integrates AM with traditional subtractive processes, offers significant advantages in precision and surface finish – but it also presents several challenges. Few of the critical challenges are as follows:

- Complexity of process integration is of the major issue, as coordinating two very different manufacturing techniques requires advanced control systems, precise calibration, and seamless software-hardware communication.
- It is observed by several researchers and practitioners that thermal stresses and material distortions get introduced during the AM stage that affects the accuracy of subsequent machining operations.
- Additionally, tool accessibility can be limited when trying to machine intricate or internal features built by AM process.
- High equipment costs, increased maintenance demands, and the need for skilled operators familiar with both AM and CNC machining further complicate adoption.
- Finally, ensuring process repeatability and quality assurance across both additive and subtractive steps remains a technical hurdle, especially when producing high-performance parts for critical applications like aerospace parts and biomedical implants.

7.9 Conclusions and Future Directions

HASM has represented a transformative advancement in modern manufacturing, combining the design flexibility and material efficiency of AM with the precision and surface finish capabilities of subtractive processes. The outlook for HASM is highly promising. However, manufacturing leaders must consider several key challenges when evaluating its suitability for their business operations. This integrated approach of additive and subtractive manufacturing addresses many limitations inherent in standalone technologies, enabling the production of complex, high-performance components with improved dimensional accuracy, surface quality, and mechanical integrity. As industries increasingly demand customized, lightweight, and functionally optimized parts, particularly in aerospace, biomedical, and automotive sectors, this hybrid manufacturing offers a viable and scalable solution. Despite current challenges related to system complexity, process integration, and cost, ongoing research and technological development continue to enhance the reliability and efficiency of these hybrid manufacturing techniques. It is expected that these technologies are going to play a critical role in the evolution of Industry 4.0, wherein a smart, adaptive and multi-functional manufacturing systems are the need of the hour.

Future research can be directed toward the following key areas:

- Extending current research attempts for developing intelligent control systems that can seamlessly integrate additive and subtractive processes including real-time monitoring, adaptive toolpath generation, and closed-loop feedback mechanisms using AI and ML.
- Another promising direction is further harnessing the potential of HASM for multi-material fabrication and functionally graded components.
- The wide adoption of HASM is currently hindered by high system complexity and capital costs, particularly among small and medium-sized enterprises. Future research should focus on developing cost-effective, modular hybrid platforms with simplified architectures and standardized interfaces.

References

[1] Vicente CMS, Sardinha M, Reis L, Ribeiro A, Leite M. Large-format additive manufacturing of polymer extrusion-based deposition systems: Review and applications, Prog Addit Manuf [Internet]. 2023;0123456789.https://doi.org/10.1007/s40964-023-00397-9

[2] Moreno Nieto D, Casal López V, Molina SI. Large-format polymeric pellet-based additive manufacturing for the Naval industry, Addit Manuf [Internet]. 2018;23:79–85. https://doi.org/10.1016/j.addma.2018.07.012

[3] Moreno Nieto D, Molina SI. Large-format fused deposition additive manufacturing: A review, Rapid Prototyp J. 2020;26(5):793–799.https://doi.org/10.1108/RPJ-05-2018-0126

[4] Goh GD, Wong KK, Tan N, Seet HL, Ling M, Nai S. Large-format additive manufacturing of polymers: A review of fabrication processes, materials, and design, Virtual Phys Prototyp [Internet]. 2024;19 (1):1–34.https://doi.org/10.1080/17452759.2024.2336160
[5] The application and development of superalloy in metal 3D printing [Internet]. 2023. https://www.eplus3d.com/the-application-and-development-of-superalloy-in-metal-3d-printing.html
[6] Deltasys e forming [Internet]. [cited 2025 Apr 9]. https://3deltasys.com/product_detail.php?name=10X
[7] Hybrid Manufacturing Technologies. Dedicated to furthering the understanding of additive manufacturing and hybrid processing [Internet]. [cited 2025 Jun 6]. https://hybridmanutech.com/resources/
[8] Rabalo MA, Rubio EM, Agustina B, Camacho AM. Hybrid additive and subtractive manufacturing: Evolution of the concept and latest trends in research and industry, Procedia CIRP [Internet]. 2023;118:741–746. https://doi.org/10.1016/j.procir.2023.06.127
[9] Wilson N, Gupta M, Mazur M, Patel MJ, Nguyen V, Gulizia S, et al. Analysis of self-supporting conformal cooling channels additively manufactured by hybrid directed energy deposition for IM tooling, Int J Adv Manuf Technol. 2024;421–441. https://doi.org/10.1007/s00170-024-13291-7
[10] Osinga T Exploring large-scale additive manufacturing [Internet]. 2024. https://www.nlr.org/wp-content/uploads/2025/01/E2021_Exploring-Large-Scale-Additive-Manufacturing.pdf
[11] Ceruti A, Liverani A, Bacciaglia A. Towards large parts manufacturing in additive technologies for aerospace and automotive applications, Procedia Comput Sci [Internet]. 2022;200:1113–1124. https://doi.org/10.1016/j.procs.2022.01.311
[12] Senwell Aerospace [Internet]. [cited 2025 Apr 9]. https://senwellexports.com/aerospace-industries
[13] CEAD Solutions [Internet]. [cited 2025 Apr 9]. https://ceadgroup.com/segments/aerospace-large-format-additive-manufacturing/
[14] Pecho P, Ažaltovič V, Kandera B, Bugaj M. Study of design and layout of UAVs 3D printed wings in relation to optimal lightweight and load distribution, Transp Res Procedia [Internet]. 2019;40:861–868. https://doi.org/10.1016/j.trpro.2019.07.121
[15] Nelson D A talk with thermwood's CEO, author of a new book on large-scale 3D printing [Internet]. 2021 [cited 2025 Apr 9]. https://www.thefabricator.com/additivereport/article/additive/a-talk-with-thermwoods-ceo-author-of-a-new-book-on-large-scale-3d-printing
[16] Vanerio D, Guagliano M, Bagherifard S. Emerging trends in large format additive manufacturing processes and hybrid techniques, Prog Addit Manuf [Internet]. 2024;0123456789.https://doi.org/10.1007/s40964-024-00771-1
[17] Feldhausen T, Paramanathan M, Heineman J, Hassen A, Heinrich L, Kurfess R, et al. Hybrid manufacturing of conformal cooling channels for tooling, J Manuf Mater Process. 2023;7(2):74. https://doi.org/10.3390/jmmp7020074
[18] Manogharan GP, Wysk RA, Harrysson OLA. Additive manufacturing-integrated hybrid manufacturing and subtractive processes: Economic model and analysis, Int J Comput Integr Manuf. 2015. https://doi.org/10.1080/0951192X.2015.1067920
[19] Mamo HB, Adamiak M, Kunwar A. 3D printed biomedical devices and their applications: A review on state-of-the-art technologies, existing challenges, and future perspectives, J Mech Behav Biomed Mater [Internet]. 2023;143:105930. https://doi.org/10.1016/j.jmbbm.2023.105930
[20] Harding A, Pramanik A, Basak AK, Prakash C, Shankar S. Application of additive manufacturing in the biomedical field: A review, Ann 3D Print Med. 2023;10:100110. https://doi.org/10.1016/j.stlm.2023.100110
[21] Salmi M. Additive manufacturing in medical engineering, J Med Eng. 2016;2016:6. http://dx.doi.org/10.1155/2016/6191526

[22] Fortune Business Insights. Hybrid additive manufacturing machine market size, share & industry analysis [Internet]. 2025 [cited 2025 Apr 9]. https://www.fortunebusinessinsights.com/hybrid-additive-manufacturing-machine-market-106123
[23] Thermwood corporation [Internet]. [cited 2025 Apr 9]. https://blog.thermwood.com/en-us/thermwood-additive-manufacturing
[24] DMG mori [Internet]. [cited 2025 Apr 9]. https://in.dmgmori.com/products/machines/lasertec?mf_workingArea=cylinder&
[25] Ingersoll machine tools [Internet]. [cited 2025 Apr 9]. https://en.machinetools.camozzi.com/products/additive-manufacturing/all-products/masterprint-.kl
[26] Phillips additive hybrid [Internet]. [cited 2025 Apr 9]. https://www.phillipscorp.com/hybrid/
[27] Matsuura corporation [Internet]. [cited 2025 Apr 9]. https://www.lumex-matsuura.com/english/lumex-avance-25
[28] Mazak introduces new hybrid metal additive manufacturing machine [Internet]. https://www.metal-am.com/mazak-introduces-new-hybrid-metal-additive-manufacturing-machine/
[29] Optomec-LENS hybrid series [Internet]. [cited 2025 Apr 9]. https://optomec.com/3d-printed-metals/lens-core-applications/hybrid-manufacturing/
[30] Diversified machine systems (DMS) [Internet]. [cited 2025 Apr 9]. https://dmscncrouters.com/product-lines/hybrid-series/
[31] Kraska T, Leitz K-H, Wolf A Method and device for laser beam welding of aluminum with reduced pore formation (European Patent EP3117982 A1). Eur Patent Off. 2017.
[32] Rahman MA, Saleh T, Jahan MP, McGarry C, Chaudhari A, Huang R, et al. Review of intelligence for additive and subtractive manufacturing: Current status and future prospects, Micromachines. 2023;14(3):508. 10.3390/mi14030508
[33] Jain R, Bharat N, Bose PSC. Digital twins of hybrid additive and subtractive manufacturing systems – A review, In Sharma VS, Dixit US, Gupta A, Verma R, and Sharma V editors. Machining and Additive Manufacturing. Lecture Notes in Mechanical Engineering. Singapore: Springer; 2024 [CPIE 2023]. https://doi.org/10.1007/978-981-99-6094-1_18
[34] Tariq U, Joy R, Wu S-H, Mahmood MA, Malik AW, Liou F. A state-of-the-art digital factory integrating digital twin for laser additive and subtractive manufacturing processes, Rapid Prototyp J. 2023;29(10):2061–2097.https://doi.org/10.1108/RPJ-03-2023-0113
[35] Raza A, Mairaj K, Russlan D, Kotiba J, Ali H, Waseem H. Incorporation of machine learning in additive manufacturing: A review, Int J Adv Manuf Technol. 2022;1143–1166.https://doi.org/10.1007/s00170-022-09916-4
[36] Iqbal A, Zhao G, Suhaimi H, He N, Hussain G, Zhao W. Readiness of subtractive and additive manufacturing and their sustainable amalgamation from the perspective of industry 4.0: A comprehensive review, Int J Adv Manuf Technol. 2020;2475–2498.https://doi.org/10.1007/s00170-020-06287-6
[37] Jayawardane H, Davies IJ, Gamage JR, John M, Biswas WK. Sustainability perspectives – a review of additive and subtractive manufacturing, Sustain Manuf Serv Econ [Internet]. 2023;2:100015. https://doi.org/10.1016/j.smse.2023.100015
[38] Breton. 3D printing and shipbuilding: the case of northern light composites' innovative sustainable hull [Internet]. 2025 [cited 2025 Apr 9]. https://breton.it/en_asia/innovation-stories/3d-printing-and-shipbuilding-the-case-of-northern-light-composites-innovative-sustainable-hull

Abdul Wahab Hashmi, Yebing Tian*, Mamilla Ravi Sankar,
Binnur Sagbas

8 Next-Generation Hybrid Post-processing of Additively Manufactured Components: Techniques, Integration, and Industry 4.0 Innovations

Abstract: Additive manufacturing (AM) enables the fabrication of complex, customized components with high design flexibility. However, limitations such as poor surface finish, dimensional inaccuracies, residual stresses, and microstructural defects often necessitate post-processing to meet industrial standards. This chapter presents a comprehensive review of hybrid post-processing strategies that synergistically integrate mechanical, thermal, chemical/electrochemical, and laser-based techniques to overcome the individual shortcomings of conventional methods, for the parts fabricated by AM. Recent advancements are highlighted, including artificial intelligence-driven parameter optimization, digital twins for predictive process control, and sustainable hybridization using eco-friendly media and energy-efficient cycles. Intra-category (e.g., machining + abrasive flow machining) and cross-category (e.g., machining + laser shock peening) combinations are analyzed in terms of surface quality (R_a < 0.1 µm), mechanical enhancement (>200% fatigue life), and process sustainability. The integration of hybrid techniques with Industry 4.0 technologies such as in situ monitoring and real-time feedback demonstrates a transformative shift toward autonomous, high-precision manufacturing. Sector-specific case studies in aerospace, biomedical, and automotive applications underscore the industrial relevance of these advancements. This chapter also provides a future-forward framework for achieving zero-defect, resource-efficient post-processing of AM parts.

Keywords: Additive manufacturing, post-processing, hybrid techniques, surface finish, dimensional accuracy, mechanical properties

8.1 Introduction

Additive manufacturing (AM) has fundamentally transformed the manufacturing landscape by enabling the production of complex, lightweight, topology-optimized structures with unprecedented geometric freedom. Unlike traditional subtractive methods, AM builds components layer by layer from digital models, facilitating mass customization, reduced material waste, and rapid prototyping of functionally graded parts [1, 2]. These capabilities have catalyzed breakthroughs in aerospace (e.g., fuel

 | https://doi.org/10.1515/9783111704456-008

nozzles with internal cooling channels), biomedical (patient-specific implants), and energy sectors (high-efficiency heat exchangers) [3, 4]. However, inherent process limitations – rooted in rapid solidification dynamics, thermal gradients, and layer-wise material deposition – compromise the functional integrity of as-fabricated components. The major challenges include:

- **Surface Irregularities**: Staircase artifacts, powder adhesion, and spatter-induced asperities elevate surface roughness ($R_a > 10 - 40\,\mu\text{m}$) [5–7].
- **Residual Stresses**: Thermal cycling induces tensile stresses (up to 80% of yield strength), promoting distortion and crack initiation [8].
- **Microstructural Defects**: Lack-of-fusion pores, keyhole porosity (5–15% volume fraction), and columnar grain growth degrade mechanical performance [9, 10].
- **Dimensional Inaccuracies**: Volumetric shrinkage and anisotropic expansion cause deviations exceeding ±200 µm [11, 12].

Post-processing methods are taxonomically classified based on their primary mechanism, as shown in Tab. 8.1 [13, 14].

Tab. 8.1: Comparison of post-processing techniques for metal AM parts.

Technique	Mechanism	Outcomes	Advantages	Disadvantages
Thermal post-processing	Heat treatments like solution heat treatment, hot isostatic pressing, and T6 to alleviate residual stresses and homogenize microstructure	Improves mechanical properties, reduces porosity, and enhances uniformity	Enhances density, improves properties, short cycle, and low energy	High temperatures/ pressures, may not fix surface defects, and grain coarsening risk
Laser peening	Short laser pulses generate plasma and pressure, inducing compressive stresses	Improves fatigue life, hardness, reduces porosity by 15–30%, and refines grains	Enhances life, surface properties, and accurate modeling	Uneven surface, needs prepolishing, and complex setup
Laser polishing	Re-melting surface with laser, redistributing material via tension and gravity	Reduces roughness (e.g., 29.3 µm to 8.4 µm) and improves hardness/ reflectivity	Automated, no bulk effect, no waste, and significant reduction	Surface-only, heat-affected zones, and parameter sensitivity
Machining and abrasive finishing	Computer numerical control milling, MAF, abrasive flow machining, microblasting for form accuracy and finish	Reduces roughness (e.g., 22.78 µm to 0.6 µm), removes defects, and boosts fatigue	Mature, accessible, and optical finishes possible	Tool wear, anisotropy issues, and potential damage

Metal AM part surface defects can be broadly categorized into geometric defects, surface irregularities, and material-related anomalies, each arising from specific thermal, deposition, or process-induced inconsistencies [15] (see Fig. 8.1).

Hybrid post-processing strategically integrates complementary techniques to overcome these trade-offs (see Fig. 8.2). By synergistically combining processes, e.g., machining (geometric correction) followed by laser shock peening (LSP) (compressive residual stress = –500 to –800 MPa), the collective efficacy surpasses individual methods [16–19]. The period 2023–2025 has witnessed transformative advancements as discussed in the following sections:

(a) **AI-Driven Process Optimization**
Machine learning algorithms now predict optimal parameters for hybrid sequences, such as:
- Convolutional neural networks (CNNs) correlate melt-pool morphology with laser polishing (LP) outcomes, reducing R_a by 38% in Ti-6Al-4V [20].
- Reinforcement learning (RL) adjusts electrochemical polishing (ECP) voltage in real time, achieving surface roughness of $R_a = 0.1$ µm on Inconel 718 [21].

(b) **Sustainable Hybridization**
- Biodegradable abrasives (e.g., walnut shells) replace alumina in abrasive flow machining (AFM), cutting waste by 30% [22–25].
- Pulse-sequenced heat treatments lower energy consumption by 25% versus conventional hot isostatic pressing (HIP) cycles [26, 27].

(c) **Industry 4.0 Integration**
- **In-Process Monitoring**: Acoustic emission sensors detect subsurface cracks during machining, triggering adaptive laser remelting [28, 29].
- **Digital Twins**: Physics-based models simulate thermal histories in hybrid processes, predicting distortion within 5% accuracy [30, 31].

(d) **Integrated AM-Subtractive Platforms**
Machines such as the DMG MORI LASERTEC 65 3D combine DED (direct energy deposition) with 5-axis computer numerical control (CNC) milling, enabling:
- On-the-fly defect correction via micromachining
- Surface finish of $R_z < 4$ µm without repositioning [32, 33]

Key post-processing techniques in AM include support removal, heat treatment, surface finishing, HIP, and coating or polishing [14, 15, 34, 35]. Each step addresses specific challenges such as eliminating supports, relieving residual stresses, improving surface texture, enhancing internal integrity, and applying functional or aesthetic coatings to ensure the final AM part meets performance requirements. Figure 8.2 presents in detail various surface post-processing or treatments applicable to metallic components/parts made by AM.

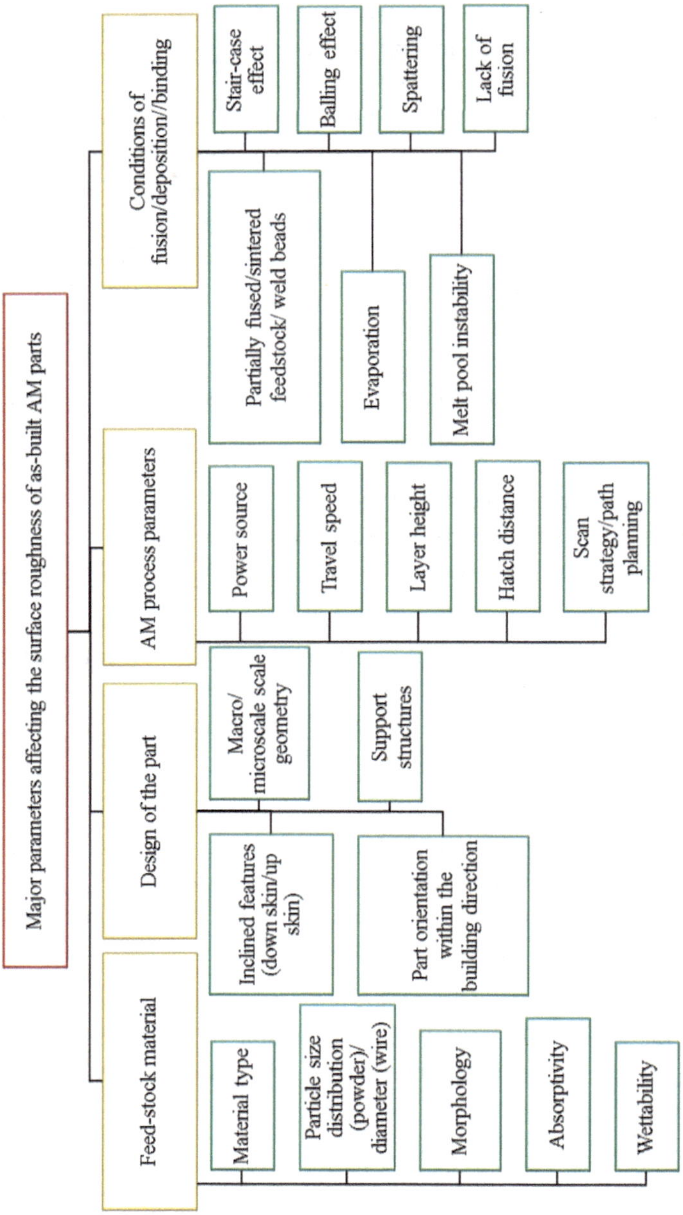

Fig. 8.1: Defect taxonomy in as-built AM parts/components [15] (with kind permission from Elsevier).

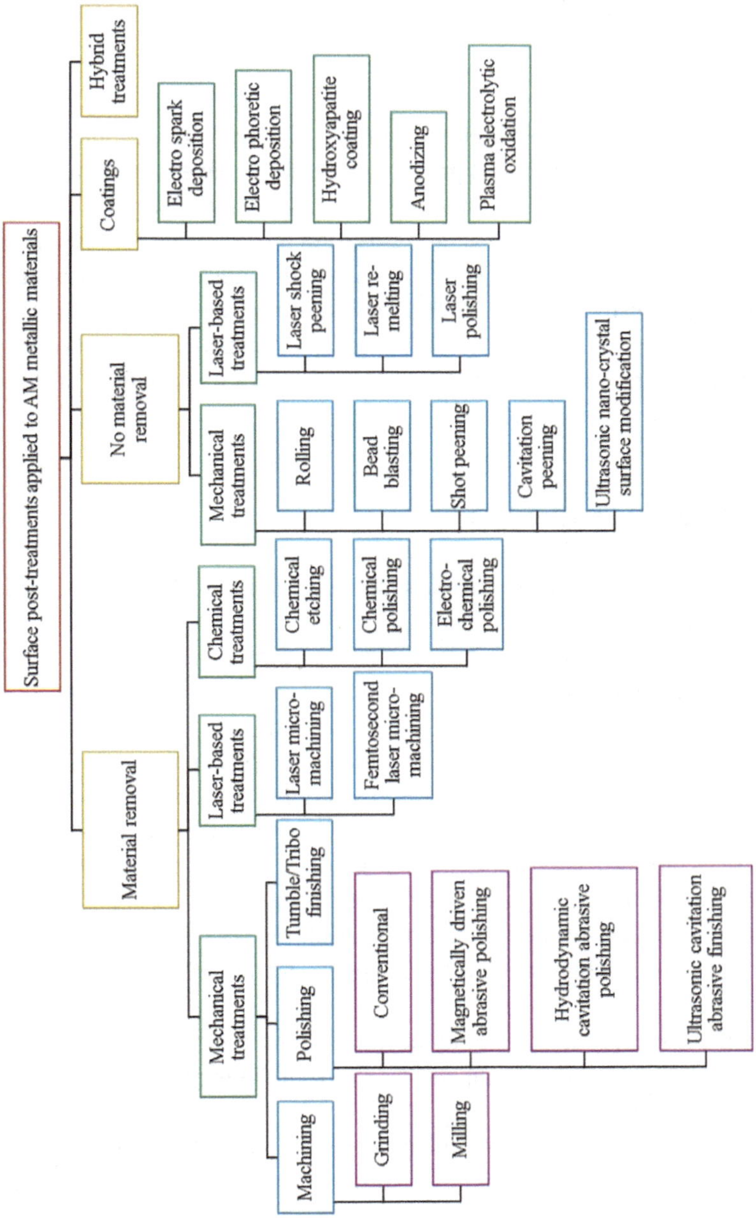

Fig. 8.2: Post-processing techniques in additive manufacturing [15] (with kind permission from Elsevier).

8.2 Fundamentals of Post-processing in Additive Manufacturing

AM components inherently exhibit suboptimal surface integrity, dimensional deviations, and heterogeneous microstructural properties due to the layer-wise deposition mechanism and rapid solidification dynamics [28, 36]. These limitations impede their deployment in fatigue-critical, tribological, and precision applications. Consequently, post-processing is indispensable to achieve compliance with industrial standards such as ISO 10110-7 (surface finish) and ASTM F3056 (mechanical properties) [37, 38]. This section delineates the root causes of surface and mechanical deficiencies in AM parts, classifies post-processing techniques, and elucidates the synergistic role of hybridization, with emphasis on advancements from 2023 to 2025.

8.2.1 Surface Quality Challenges

The surface topography of as-built AM parts is characterized by stochastic irregularities stemming from process-specific artifacts:

- **Staircase Effect**: Layer thickness (typically 20–100 µm) and build orientation (θ) induce step-like deviations, quantified by the average roughness parameter $R_a \propto \text{layer thickness} \times |\cos(\theta)|/(1 - |\cos(\theta)|)$ [39–41].
- **Powder Adhesion**: Partially melted particles (diameter: 15–45 µm) adhere to surfaces, exacerbating roughness in powder bed fusion (PBF) processes. For instance, selective laser melting (SLM) Ti-6Al-4V exhibits R_a values of 10–25 µm [42–44].
- **Thermal Stress-Induced Warping**: Thermal gradients ($\nabla T > 10^3$ K/mm) during solidification promote surface buckling and balling phenomena [45, 46].
- **Subsurface Defects**: Lack-of-fusion pores and gas-entrapped keyholes (diameter: 20–200 µm) concentrate stresses, reducing fatigue strength by 30–50% [10, 47, 48].

LP significantly enhances the surface quality of AM parts by removing adhered particles and reducing surface roughness [49], as evident from the improved topography of SLM Inconel 718 after treatment.

Hybrid LP, combining pulsed and continuous wave (CW) laser methods, enhances surface quality of AM parts. Figure 8.3 illustrates: (a) the experimental setup for pulsed and CW LP, (b) the scanning trajectory of hybrid LP, (c) the pulsed LP process schematic, and (d) the CW LP process schematic, demonstrating uniform energy distribution for reduced roughness and improved microtexture [50].

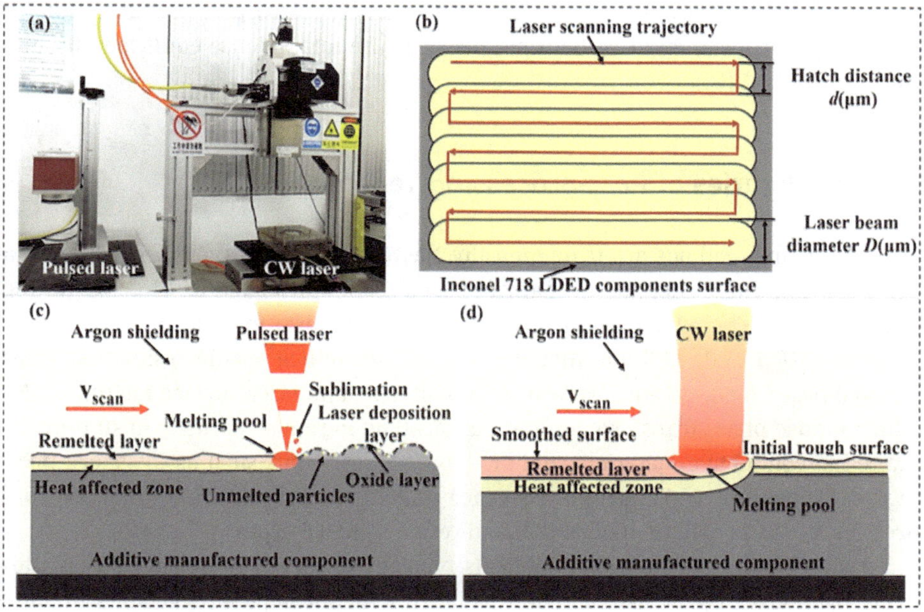

Fig. 8.3: Hybrid laser polishing of AM parts [50] (with kind permission from Elsevier).

8.2.2 Limitations of Mechanical Properties

The mechanical performance of AM parts is constrained by defects originating from metastable solidification:

- **Porosity**: Volumetric pore fractions of 0.5–5% (PBF) and up to 15% (directed energy deposition) act as stress concentrators, reducing fatigue strength by 40–60% compared to wrought counterparts [51, 52].
- **Residual Stresses**: Tensile residual stresses (200–500 MPa) develop due to thermal contraction constraints, diminishing fracture toughness by 15–30% [53–55].
- **Anisotropic Microstructure**: Columnar grain growth along the build direction and heterogeneous phase distribution yield directionally dependent properties (e.g., 20% lower transverse ductility) [56, 57].

Thermomechanical post-processing techniques such as HIP and LSP mitigate these defects:

- **HIP**: Applying isostatic pressure (100–200 MPa) at 0.6–0.9 T_m (melting temperature) collapses pores and homogenizes microstructure [58, 59].
- **LSP**: Laser-induced plasma shocks ($P > 5$ GPa) generate compressive residual stresses (depth: 0.5–1 mm, magnitude: −800 MPa), enhancing fatigue life by three to five times [60–62].

HIP effectively reduces internal porosity and enhances microstructural uniformity in AM parts, transforming the as-built porous structure into a dense equiaxed phase, as shown in Fig. 8.4 [63].

8.2.3 Categories of Post-processing Techniques

Post-processing methods are taxonomically classified based on their primary mechanism [14, 15]. These include mechanical-based techniques such as CNC machining, grinding, and AFM, which remove material via mechanical force to improve surface roughness (R_a) to 0.4–1.6 µm, though they may introduce tensile residual stresses; thermal-based methods such as heat treatment, HIP, and laser surface melting (LSM), which induce phase transformation or viscous flow to reduce porosity to <0.1 µm but limited to conductive materials; and laser-based treatments such as LP, shock peening, and remelting, which generate surface remelting or shockwaves to induce compressive stresses (−500 to −1,000 MPa) and refine R_a to 0.8–2 µm.

A classification of multi-scale defects typically observed in powder-based additively manufactured metals and alloys is illustrated, encompassing geometry-related issues such as part distortion and delamination, surface integrity concerns including the staircase effect, partially melted powder, balling, and surface cracking, as well as microstructural defects like internal cracks, various pore types, and textured columnar grains [64]. As illustrated in Fig. 8.5, these defects span from millimeter to nanometer scales and significantly affect the performance of AM parts.

8.2.4 Role of Hybridization

Hybrid post-processing integrates complementary techniques to concurrently address multiple deficiencies. The synergistic effects arise from sequential or simultaneous application, as discussed below:

- **Mechanical-Thermal Hybrids**: Machining followed by HIP enhances dimensional accuracy while eliminating subsurface pores. For Ti-6Al-4V, this sequence improves fatigue strength by 120% versus standalone machining [65, 66].
- **Laser-Chemical Hybrids**: LSM homogenizes the near-surface microstructure, followed by electropolishing (EP) to achieve optical-grade finishes ($R_a < 0.05$ µm) [67–69].
- **AI-Optimized Hybrids**: RL algorithms dynamically adjust parameters (e.g., AFM pressure and LSP pulse count) based on real-time surface metrology, reducing R_a by 45% in Inconel 718 [70, 71].

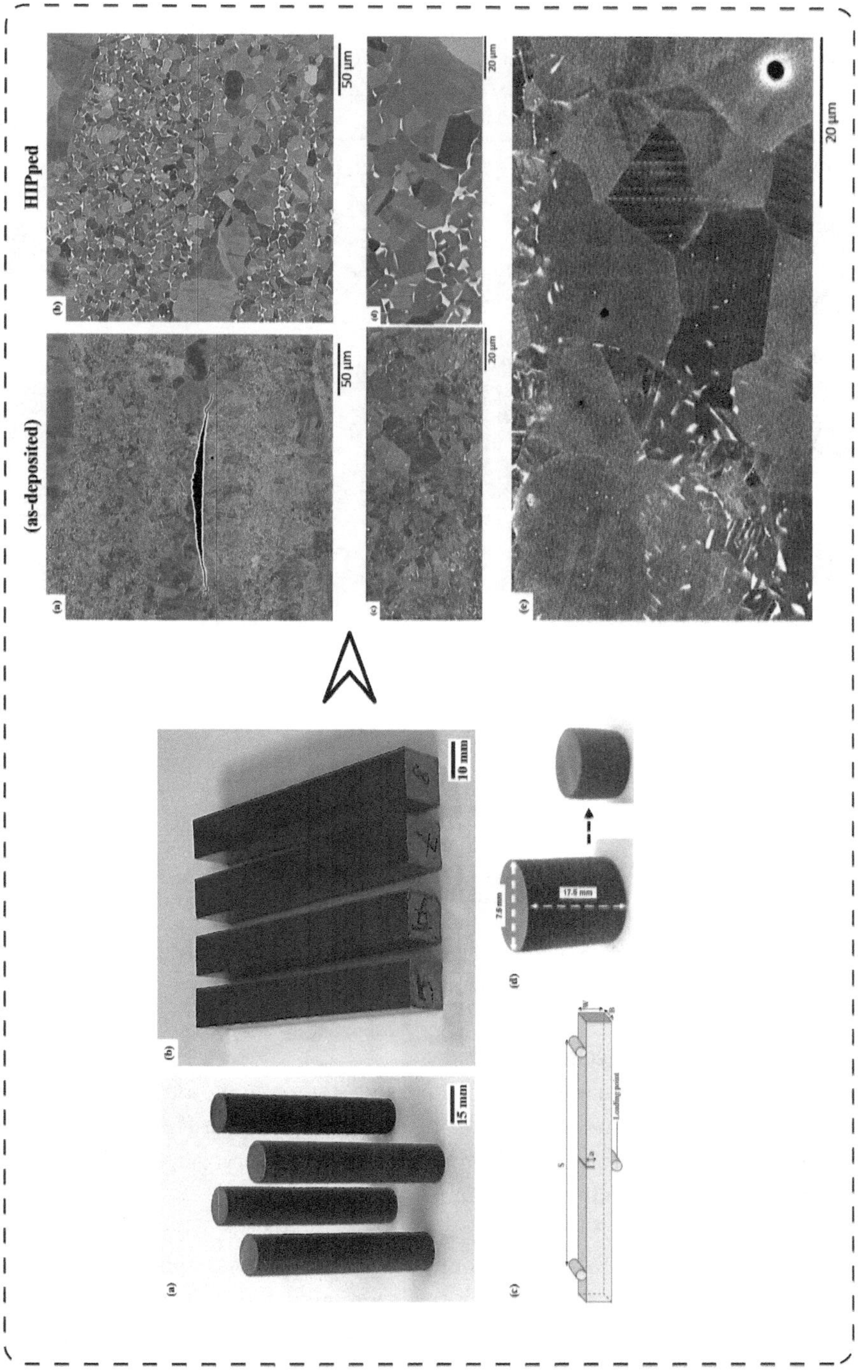

Fig. 8.4: Results of microstructural analysis of HIP-treated AM parts [63] (with kind permission from Elsevier).

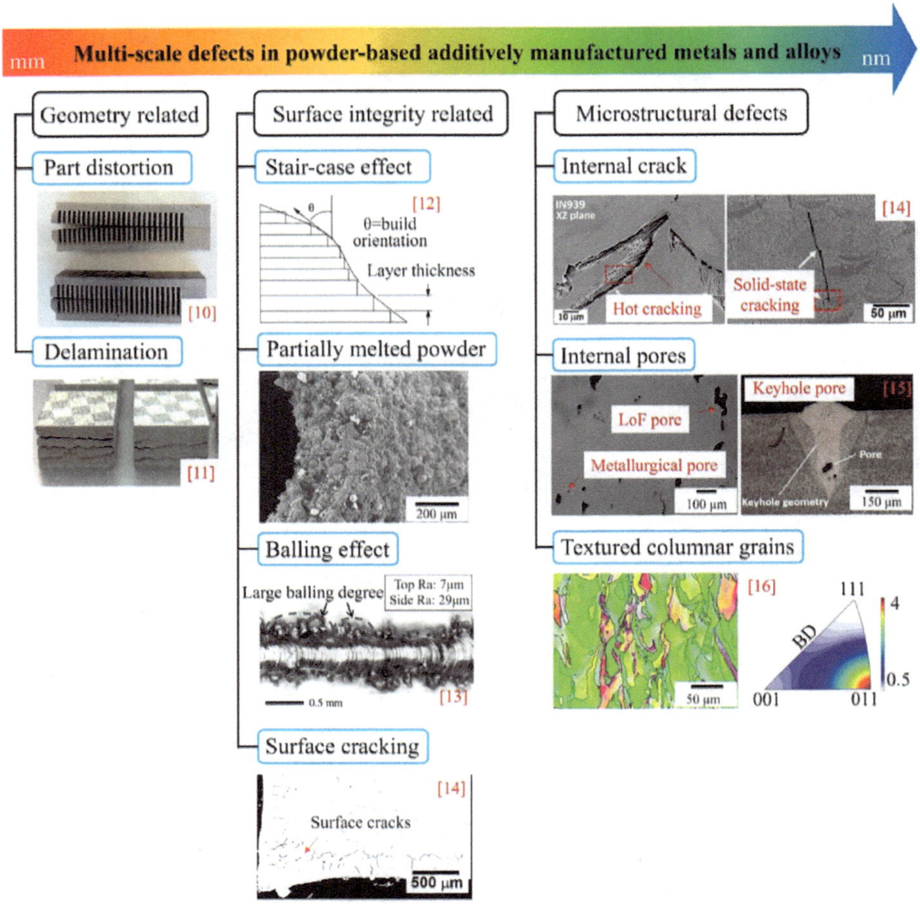

Fig. 8.5: Overview of geometry, surface, and microstructure-related defects in powder-based metal AM components across different length scales [64] (with kind permission from Elsevier).

The recent (2024–2025) advancements include:

- **Closed-Loop Systems**: In situ optical coherence tomography during LP enables adaptive focal spot adjustment, minimizing heat-affected zones [72–74].
- **Sustainable Hybrids**: Water jet machining coupled with cryogenic cooling eliminates cutting fluids, reducing environmental impact [75–77].

The post-processing of metal AM parts follows a systematic sequence aimed at addressing inherent limitations such as poor surface quality and internal defects. This includes applying surface finishing techniques, mechanical treatments, and thermal methods like HIP and heat treatment to enhance surface integrity and mechanical performance. The integration of AI-driven post-processing aligned with Industry 4.0 further enables intelligent and adaptive control for high-performance applications.

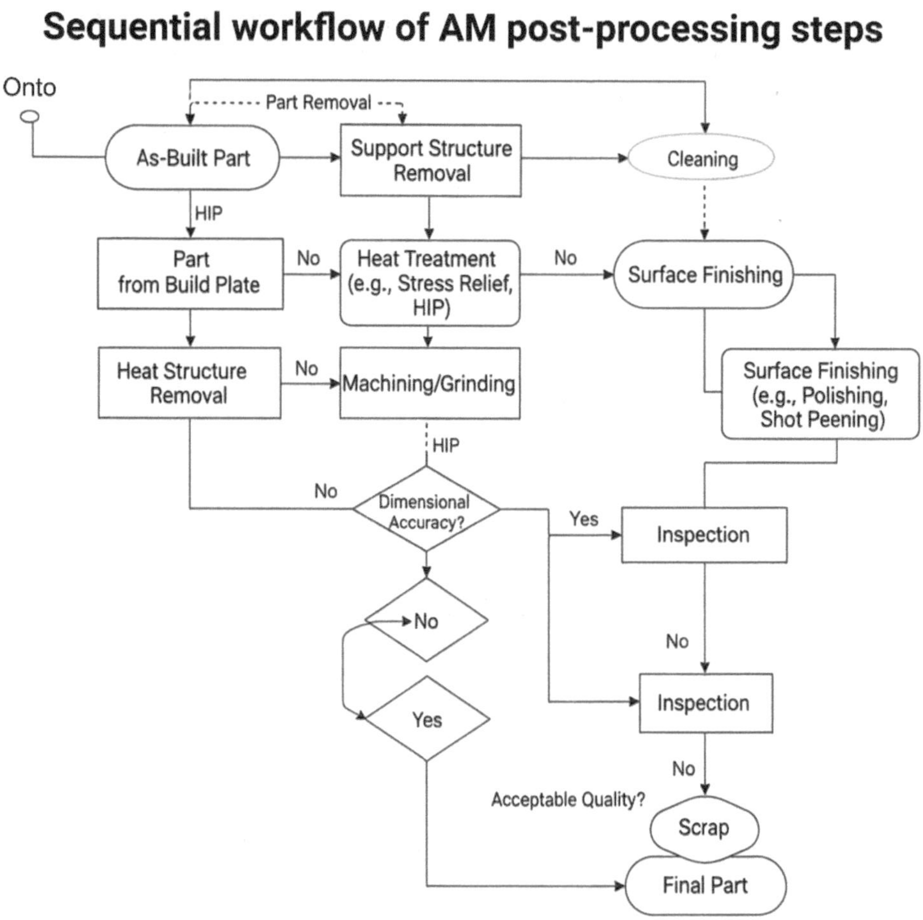

Fig. 8.6: Sequential workflow of AM post-processing steps.

Figure 8.6 illustrates the sequential workflow of post-processing steps for AM parts until the desirable quality is obtained.

8.3 Hybrid Post-processing Techniques

Hybrid post-processing synergistically integrates multiple techniques to overcome the inherent limitations of individual methods, achieving unprecedented improvements in surface finish (R_a = 0.1 µm), dimensional accuracy (IT grade 4–5), and mechanical performance (fatigue life enhancement >200%) [78–80]. This section systematically examines intra-category (homogeneous techniques) and cross-category (heterogeneous techniques) hybrid approaches, emphasizing mechanistic synergies, parametric interdependencies, and industrial applications validated by recent studies (2023–2025).

Hybrid post-processing techniques for AM parts are categorized based on the synergy of combined methods to overcome individual limitations. As presented in Fig. 8.7, these include mechanical-thermal (e.g., AFM + heat treatment) for enhancing surface and internal properties, mechanical-chemical for improving precision and finish, laser-chemical for defect reduction and microstructural refinement, and multimodal combinations that integrate several techniques for comprehensive part enhancement.

8.3.1 Intra-category Hybrid Techniques

8.3.1.1 Mechanical-Mechanical Hybrid Techniques

Combining subtractive machining with AFM addresses both macroscale geometric inaccuracies and microscale surface asperities:

- **Synergistic Mechanism**: Machining (e.g., micromilling) corrects dimensional deviations (±15 μm), while AFM's viscoelastic abrasive media (SiC/AlO) induces uniform plastic deformation in internal features [81, 82].
- **Example**: For Ti-6Al-4V aerospace impellers, sequential machining (end milling) and AFM (media pressure: 8 MPa) reduced R_a from 18.3 μm to 3.7 μm (80% reduction) and increased high-cycle fatigue strength from 325 MPa to 580 MPa [83].
- **Recent Advances**: RL optimizes AFM parameters (abrasive size and viscosity) in real time, cutting processing time by 35% [84, 85].

8.3.1.2 Thermal-Thermal Techniques

Sequential solution annealing and HIP eliminate defects in precipitation-hardened superalloys:

- **Synergistic Mechanism**: Solution treatment (1,050 °C/1 h) dissolves γ′ phases, while HIP (1,180 °C/100 MPa/4 h) collapses porosity via creep diffusion [86, 87].
- **Quantitative Outcomes**: For Inconel 718, this hybrid reduced porosity from 0.8% to <0.01% and increased stress rupture life by 3.2× at 650 °C [88, 89].
- **AI Integration**: Neural networks predict optimal HIP parameters, reducing energy consumption by 22% [90, 91].

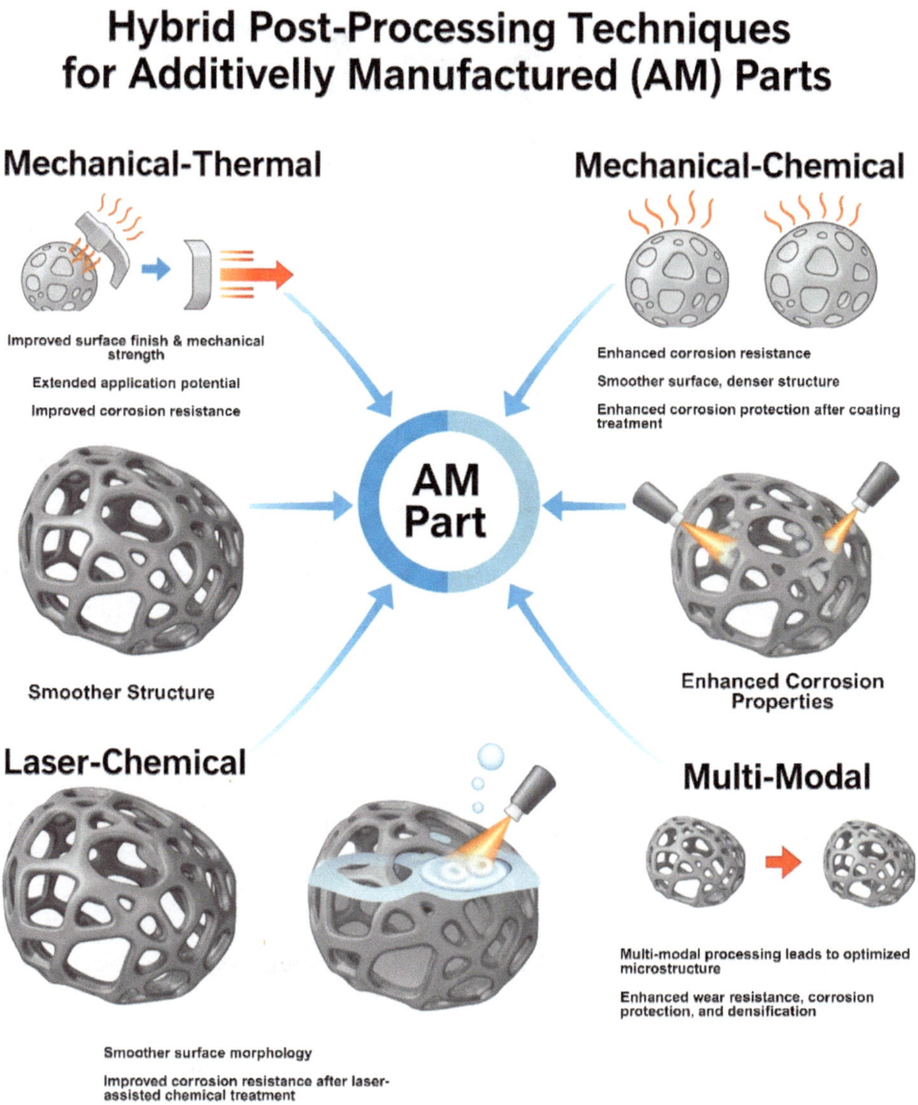

Fig. 8.7: Various hybrid post-processing techniques for AM parts.

8.3.1.3 Chemical-Electrochemical Hybrid Techniques

EP with chemical etching (CE) achieves nanoscale surface finishes:

– **Synergistic Mechanism**: EP (voltage: 12 V), electrolyte: H_2SO_4 – H_3PO_4 removes bulk irregularities, while CE (HF–HNO_3) refines microtextures via isotropic dissolution [92, 93].

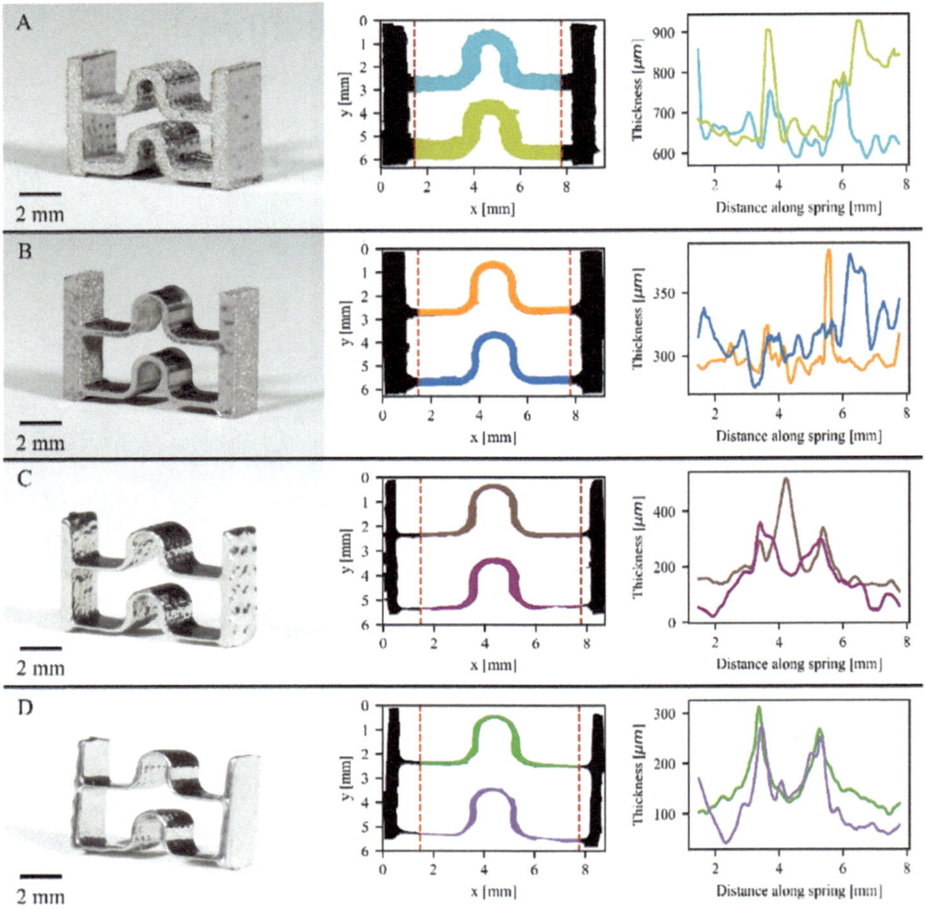

Fig. 8.8: Results of post-processing of NiTi actuators fabricated by AM [97]. From *Additive Manufacturing Letters* (Elsevier), used under Creative Commons Attribution (CC BY) 4.0 License.

- **Sustainability Innovation**: Biodegradable electrolytes (citric acid-based) reduced toxic sludge generation by 30% for 316L stainless steel [94, 95].
- **Efficacy**: R_a reduction from 8.4 µm to 0.07 µm in conformal cooling channels [96].

Figure 8.8 presents a comparative assessment of NiTi AM actuator samples before and after post-processing, highlighting surface improvements and dimensional accuracy. It is the best example of this type of hybrid process. As-printed (A) surfaces show high roughness and oversize thickness, whereas machined (B) and chemically electropolished (C, D) variants reveal smoother finishes and convergence toward target thickness (~300 µm) [97], especially in the CE10 + EP12 condition.

8.3.1.4 Laser-Laser Hybrid Techniques

LP followed by LSP enhances surface integrity as per the following:
- **Synergistic Mechanism**: LP (power: 300 W, scan speed: 500 mm/s) remelts surface layers (depth: 50–100 μm), while LSP (pulse energy: 8 J, spot size: 3 mm) induces compressive stresses (−700 MPa) [98–100].
- **Fatigue Enhancement**: For Ti-6Al-4V, LP + LSP increased fatigue strength from 450 MPa to 810 MPa [101].
- **Emerging Trend**: Ultrafast femtosecond lasers reduced HAZ thickness by 40% in AlSi10Mg [102].

8.3.2 Cross-Category Hybrid Techniques

8.3.2.1 Mechanical-Thermal Hybrid Techniques

Machining and heat treatment synergistically enhance precision and microstructural homogeneity:
- **Mechanism**: Finish machining (tolerance: ±5 μm) relieves surface stresses, while aging (480 °C/6 h) precipitates Ni_3Ti phases in maraging steel [103, 104].
- **Case Study**: Automotive gears showed 90% reduction in distortion and hardness increase to 54 HRC [105].
- **Process Flow**: Adaptive sequencing (machining → aging → finish machining) minimizes distortion [106].

8.3.2.2 Mechanical-Laser Hybrid Techniques

Machining combined with LSP enhances fatigue resistance with the following details:
- **Mechanism**: Milling achieves $R_a < 1.6$ μm, while LSP introduces compressive residual stresses (−500 MPa) to inhibit crack initiation [107, 108].
- **AI-Driven Optimization**: Real-time acoustic emission feedback adjusts LSP pulse density, improving fatigue life by 35% in turbine blades [109].

8.3.2.3 Thermal-Chemical Hybrid Techniques

HIP and EP eliminate defects and improve surface finish:
- **Mechanism**: HIP (920 °C/100 MPa) closes internal pores, while EP removes oxide layers without mechanical deformation [110, 111].
- **Sustainability**: Pulse-sequenced HIP cycles reduced energy use by 28%, while non-toxic electrolytes (choline chloride-based) decreased wastewater toxicity [112].

8.3.2.4 Laser-Chemical Hybrid Techniques

LSM with ECP achieves optical-grade finishes. The relevant details are:
- **Mechanism**: LSM (power density: 10^5 W/cm) homogenizes microstructure, while ECP (current density: 0.5 A/cm) eliminates microripples [113, 114].
- **Innovation**: In situ hyperspectral monitoring adjusts ECP voltage based on oxide layer thickness, achieving R_a = 0.05 μm [115].

8.4 Process Parameters and Optimization in Hybrid Post-processing

The synergistic benefits of hybrid post-processing are contingent upon precise optimization of process parameters and sequencing strategies. These variables govern surface integrity (R_a dimensional tolerances (IT grades), mechanical properties (fatigue strength and hardness), and sustainability metrics (energy consumption and waste) [116, 117]. The following sections employ a systems-engineering approach to analyze parametric interdependencies, sequencing effects, and AI-enabled optimization frameworks, with emphasis on 2024–2025 advancements.

8.4.1 Parametric Influence on Hybrid Outcomes

The key parameters exhibit nonlinear relationships with output characteristics as shown in Tab. 8.2.

8.4.2 Sequencing Impact on Dimensional Stability

Process sequencing in hybrid post-processing of additively manufactured components significantly impacts residual stress and geometric accuracy, with techniques like mechanical-thermal, thermal-mechanical, and laser-chemical sequencing addressing specific challenges. Mechanical-thermal sequencing minimizes distortion in maraging steel to under 5 μm/m but risks microcracking [119], while thermal-mechanical sequencing, using HIP before machining, reduces porosity but increases tool wear by 15–25% due to work hardening [120]. Laser-chemical sequencing enhances surface finish by removing oxide layers, improving current efficiency by 35% [121]. Recent advancements (2023–2025) leverage AI-driven CNNs to optimize laser parameters, reducing roughness by 40% [122], and digital twins for predictive maintenance with 80% accuracy [123], enabling tailored solutions for aerospace and biomedical applications [124].

Tab. 8.2: Key process parameters, mechanisms, and performance gains in AM post-processing techniques [14, 15, 118].

Parameter	Influence mechanism	Optimal range	Performance gain
Laser power density (W/cm^2)	Determines melt depth in laser polishing; excessive power induces vaporization pits	10^4–10^5	R_a reduction: 50–80%
AFM media pressure (MPa)	Governs abrasive embedment depth; high pressure causes work hardening	5–8 (metals), 2–4 (polymers)	Internal R_a < 1.5 μm
Electropolishing current density (A/cm^2)	Controls dissolution uniformity; low density causes pitting	0.3–0.7	Mirror finishes (R_a < 0.05 μm)
HIP temperature (°C)	Dictates diffusional pore closure; exceeding β-transus causes grain growth	0.85–0.95 T_m	Porosity reduction: >99%

8.5 Advanced Hybrid Technologies and Industry 4.0 Integration

Advanced hybrid technologies in AM post-processing, integrated with Industry 4.0 principles, combine additive-subtractive platforms, artificial intelligence (AI) optimizations, and digital twins to enhance precision, efficiency, and sustainability. The integration of AI, physics-based modeling, and digital twins revolutionizes hybrid post-processing for additively manufactured components, enhancing precision and efficiency. CNNs analyze surface topology to optimize laser parameters, reducing surface roughness (R_a) by 52% in fewer than 10 iterations [125], while RL adjusts AFM viscosity in real time, cutting cycle times by 40% [126]. Physics-based simulations, such as finite element analysis with Johnson-Cook plasticity, predict LSP stresses with less than 8% error [127], and computational fluid dynamics eliminates dead zones during EP by simulating electrolyte flow [128]. Digital twins, leveraging Internet of Things (IoT) sensors such as acoustic emission and pyrometers, enable real-time adaptive control, correcting laser focal distance during polishing [129] and providing predictive maintenance for HIP furnaces with 92% accuracy [130].

Hybrid AM (HAM) systems integrate directed energy deposition with CNC machining for in situ post-processing, addressing assembly and finishing challenges while reducing material waste [131]. The seamless incorporation of Industry 4.0 elements, such as IoT, big data, and cyber-physical systems, enables real-time monitoring, predictive analytics, and adaptive control in AM workflows [132]. Recent developments (2023–2025) emphasize automated electrochemical post-processing techniques, like Hirtisation, for surface refinement and database-driven characterization of AM parts [133]. Digital twins facilitate multi-physics simulations for process optimization, significantly lowering scrap rates and energy use in hybrid manufacturing cells [134]. Furthermore, the expansion of 3D printing within Industry 4.0 supports agile, customized production across sectors such as aerospace and healthcare, fostering supply chain resilience [135]. Market projections highlight robust growth in hybrid AM technologies, driven by innovations in materials and applications through 2032 [136].

8.6 Conclusion and Future Research Directions

8.6.1 Conclusion

Hybrid post-processing represents a transformative paradigm in AM, elevating as-built components from mere prototypes to high-performance industrial assets by synergistically integrating mechanical, thermal, electrochemical, and laser-based techniques to overcome inherent limitations in surface finish, dimensional accuracy, and microstructural integrity. Through innovative hybrid sequences, such as machining combined with HIP and laser peening, these approaches concurrently eradicate porosity to negligible levels, impart deep compressive residual stresses, and yield mirror-like surfaces, thereby dramatically boosting fatigue endurance and operational reliability of AM fabricated parts across demanding sectors such as aerospace, biomedical, and energy. The infusion of AI, including CNNs and RL, enables real-time parameter optimization that slashes surface roughness and energy demands, while physics-informed models forecast stress distributions with exceptional fidelity to facilitate distortion-free outcomes. Embracing sustainability, these strategies incorporate eco-friendly abrasives, benign electrolytes, and efficient thermal pulsing to curtail environmental footprints, alongside closed-loop recycling systems that maximize material reuse and minimize waste. Finally, the seamless fusion with Industry 4.0 technologies encompassing digital twins, advanced in situ monitoring, intelligent process control, and optimization ultimately unlocks unprecedented customization, efficiency, and innovation in the production of complex parts by AM.

8.6.2 Future Research Directions

To address many challenges and fulfill the requirements, the following future research avenues are identified:

- **Multi-scale Process Modeling**: Coupling molecular dynamics with continuum models to predict phase transformations in hybrid thermal cycles.
- **Cyber-Physical Security**: Blockchain frameworks for IP protection in distributed hybrid manufacturing for AM parts.

- **Bio-inspired Hybrids**: Biomimetic surface texturing via combined laser-electrochemical processing for enhanced tribology of parts fabricated by AM.
- **Combining Intelligence and Sustainability**: Optimization of hybrid techniques to attain sustainability in AM post-processing.

Hybrid post-processing has evolved from a corrective step to a value-generating pillar of AM, enabling high-stakes applications in aerospace, biomedicine, and energy. As AI, IoT, and sustainable design converge, this field will catalyze the next leap toward autonomous, eco-efficient manufacturing.

Acknowledgments

This work was financially supported by the National Natural Science Foundation of China (Nos. 52575516, 51875329), Taishan Scholar Special Foundation of Shandong Province (Nos. tstp20240826, tsqn201812064), Ningxia Hui Autonomous Region Key Research and Development Project (No. 2024BEE02019), and Shandong Provincial Natural Science Foundation (No. ZR2023ME112).

References

[1] Al Ali M. Advanced techniques in porous structure design for additive manufacturing. Hoboken (NJ): John Wiley & Sons; 2025.
[2] Maniraj M, Kumar SR. Design for additive manufacturing: bridging creativity and precision in the 3D printing era, In Modeling, Analysis, and Control of 3D Printing Processes. Hershey (PA): IGI Global Scientific Publishing; 2025. p. 79–100.
[3] Fidan I, Imeri A, Gupta A, Hasanov S, Nasirov A, Elliott A. et al. Recent inventions in additive manufacturing: Holistic review, Inventions. 2023;8(4):103.
[4] Bose S, Sarkar S, Banerjee D, Bandyopadhyay A. 3D printing of ceramics: Advantages, challenges, applications, and perspectives, J Am Ceram Soc. 2024;107(12):7879–7920.
[5] Zhang H, Vallabh CKP, Zhao X. Influence of spattering on in-process layer surface roughness during laser powder bed fusion, J Manuf Process. 2023;104:289–306.
[6] Hartmann C, Van den Bosch L, Spiegel J, Rumschöttel D, Günther D. Removal of stair-step effects in binder jetting additive manufacturing using grayscale and dithering-based droplet distribution, Materials. 2022;15(11):3798.
[7] Bahnini I, Rivette M, Rechia A, Siadat A, Elmesbahi A. Additive manufacturing technology: The status, applications, and prospects, Int J Adv Manuf Technol. 2018;97(1):147–161.
[8] Li Y, Wang J, Wang X. Improvement of stress-relief cracking resistance in coarse- grained heat-affected zone of T23 steel by refining sub-structure through second thermal cycle, J Mater Res Technol. 2020;9(4):8568–8579.
[9] Kan WH, Hoang LS, Karimi P, Su Y, Panwisawas C, Leung CLK. et al. A critical review on the effects of process-induced porosity on the mechanical properties of alloys fabricated by laser powder bed fusion, J Mater Sci. 2022;57(21):9818–9865.

[10] Brennan MC, Keist JS, Palmer TA. Defects in metal additive manufacturing processes, J Mater Eng Perform. 2021;30:4808–4818. https://doi.org/10.1007/s11665-021-05919-6.

[11] Tantbirojn D, Pfeifer CS, Amini AN, Versluis A. Simple optical method for measuring free shrinkage, Dent Mater. 2015;31(11):1271–1278.

[12] Zago M, Lecis NFM, Vedani M, Cristofolini I. Dimensional and geometrical precision of parts produced by binder jetting process as affected by the anisotropic shrinkage on sintering, Addit Manuf. 2021;43:102007.

[13] Hashmi AW, Mali HS, Meena A. Improving the surface characteristics of additively manufactured parts: A review, Mater Today Proc. 2023;81:723–738.

[14] Hashmi AW, Mali HS, Meena A. A comprehensive review on surface quality improvement methods for additively manufactured parts, Rapid Prototyp J. 2023;29(3):504–557. doi:10.1108/RPJ-06-2021-0133.

[15] Maleki E, Bagherifard S, Bandini M, Guagliano M. Surface post-treatments for metal additive manufacturing: progress, challenges, and opportunities, Addit Manuf. 2021;37:101619. 10.1016/j.addma.2020.101619.

[16] Ali S, Kim D. Enhanced power control model based on hybrid prediction and preprocessing/post-processing, J Intell Fuzzy Syst. 2016;30(6):3399–3410.

[17] Wang G, Ascheid G. Joint pre/post-processing design for large millimeter wave hybrid spatial processing systems, In Proceedings of the European Wireless 2014; 20th European Wireless Conference. VDE. 2014. p. 1–6.

[18] Wakchaure MB, Misra M, Menezes PL. A comprehensive review on finite element analysis of laser shock peening, Materials. 2024;17(17):4174.

[19] Madireddy GCR. Modeling Residual Stress Development in Hybrid Processing by Additive Manufacturing and Laser Shock Peening. Master's thesis. Lincoln (NE): University of Nebraska-Lincoln; 2018.

[20] Lee H. Convolutional Neural Network-based Pore Detection using the Melt Pool Monitoring Data in the Selective Laser Melting Additive Manufacturing Process. Master's thesis. Seoul: Seoul National University; 2022.

[21] Rajput AS, Das M, Kapil S. Machine-learning-based optimization of hybrid electrochemical magnetorheological finishing process to achieve nano finishing on additively manufactured biomaterial, Adv Mater Process Technol, 2025;1–20.

[22] Karkalos NE, Karmiris-Obratański P. Determination of the feasibility of using eco- friendly walnut shell abrasive particles for pocket milling of titanium workpieces by abrasive waterjet technology, Metals. 2023;13(10):1645.

[23] Hashmi AW, Meena A, Mali HS, Yadav P, Kumar M. et al. Abrasive flow finishing of FDM printed extrusion die insert pattern using novel AFM fixture with mandrel guide, Surf Rev Lett. 2023;30 (6):2350034.

[24] Hashmi AW, Mali HS, Meena A, Ahmad S, Tian Y.. A novel eco-friendly abrasive media-based abrasive flow machining of 3D printed PLA parts using IGWO and ANN, Rapid Prototyp J. 2023; [Epub ahead of print].

[25] Hashmi AW, Mali HS, Meena A, Ahmad S, Tian Y. et al. Abrasive flow finishing of 3D-printed Aerofoils: Design, numerical simulation, and experimental analysis, Opt Laser Technol. 2024;174:110578.

[26] Careri F, Imbrogno S, Umbrello D, Attallah MM, Outeiro J, Batista AC. Machining and heat treatment as post-processing strategies for Ni-superalloy structures fabricated using direct energy deposition, J Manuf Process. 2021;61:236–244.

[27] Li P, Gong Y, Liang C, Yang Y, Cai M Effect of post-heat treatment on residual stress and tensile strength of hybrid additive and subtractive manufacturing, Int J Adv Manuf Technol. 2019;103 (5):2579–2592.

[28] Chen L. Multi-sensor Monitoring for In-situ Defect Detection and Quality Assurance in Laser-directed Energy Deposition. Doctoral thesis. Singapore: Nanyang Technological University; 2024.

[29] Donmez A, Slotwinski J, Moylan S, Fesperman R. et al. In-process monitoring and non-destructive evaluation for metal additive manufacturing processes. Gaithersburg (MD): US Department of Commerce, National Institute of Standards and Technology; 2024.

[30] Pawar S, Ahmed SE, San O, Rasheed A. Hybrid analysis and modeling for next generation of digital twins, In Journal of Physics: Conference Series. Bristol (UK): IOP Publishing; 2021. p. 012031.

[31] Di Meglio A, Massarotti N, Nithiarasu P. A physics-driven and machine learning- based digital twinning approach to transient thermal systems, Int J Numer Methods Heat Fluid Flow. 2024;34 (6):2229–2256.

[32] Oyelola O. Machining of Metal Directed Energy Deposited Components. Doctoral thesis. Nottingham: University of Nottingham; 2020.

[33] Dugar J, Ikram A, Klobčar D, Pušavec F. Sustainable hybrid manufacturing of alsi5 alloy turbine blade prototype by robotic direct energy layered deposition and subsequent milling: an alternative to selective laser melting?, Materials. 2022;15(23):8631.

[34] Kantaros A, Ganetsos T, Petrescu FIT, Ungureanu LM, Munteanu IS. Post-production finishing processes utilized in 3D printing technologies, Processes. 2024;12(3):595.

[35] Mahmood MA, Chioibasu D, Ur Rehman A, Mihai S, Popescu AC. Post-processing techniques to enhance the quality of metallic parts produced by additive manufacturing, Metals. 2022;12(1):77.

[36] Karthik G, Kim HS. Heterogeneous aspects of additive manufactured metallic parts: A review, Met Mater Int. 2021;27(1):1–39.

[37] Elambasseril J, Rogers J, Wallbrink C, Munk D, Leary M, Qian M. Laser Powder Bed Fusion Additive Manufacturing (LPBF-AM): The influence of design features and LPBF Variables On Surface Topography And Effect On Fatigue Properties, Crit Rev Solid State Mater Sci. 2023;48(1):132–168.

[38] Zhang Q, Zuo Y, Hu Z, Xu Y, Liu X. Advances and challenges in interference-fit technology for enhancing the mechanical performance of joints, J Mater Eng Perform. 2025;34(5):3585–3607.

[39] Krishna AV. On characterization and optimization of surface topography in additive manufacturing processes. Gothenburg (Sweden): Chalmers Tekniska Högskola; 2022.

[40] Lesyk D, Martinez S, Mordyuk B, Dzhemelinskyi V, Lamikiz A, Prokopenko G. Post-processing of the Inconel 718 alloy parts fabricated by selective laser melting: Effects of mechanical surface treatments on surface topography, porosity, hardness and residual stress, Surf Coat Technol. 2020;381:125136.

[41] Feng S, Chen S, Kamat AM, Zhang R, Huang M, Hu L. Investigation on shape deviation of horizontal interior circular channels fabricated by laser powder bed fusion, Addit Manuf. 2020;36:101585.

[42] Mekhiel S. Surface Modification of Additively Manufactured Metallic Components. Doctoral thesis. Hamilton (ON): McMaster University; 2021.

[43] Carter LN, Villapún VM, Grover L, Cox SC. Exploring the duality of powder adhesion and underlying surface roughness in laser powder bed fusion processed Ti- 6Al-4V, J Manuf Process. 2022;81:14–26.

[44] Yousefi M. The Role of Post Surface Finishing on the Surface Quality and Fatigue Behavior of Ti-6Al-4V Alloy Produced via Laser Powder Bed Fusion Additive Manufacturing Process. Master's thesis. Turin: Politecnico di Torino; 2024.

[45] Ling W, Li X, Xu Z, Zhang Z, Wei Y. Thermal effects of fiber sensing coils in different winding patterns considering both thermal gradient and thermal stress, Opt Commun. 2015;356:290–295.

[46] Klingbeil NW, Beuth JL, Chin R, Amon C. Residual stress-induced warping in direct metal solid freeform fabrication, Int J Mech Sci. 2002;44(1):57–77.

[47] Yang KV, Rometsch P, Davies C, Wu X, Huang A. Porosity formation mechanisms and fatigue response in Al-Si-Mg alloys made by selective laser melting, Mater Sci Eng A. 2018;712:166–174.

[48] Al-Maharma AY, Patil SP, Markert B. Effects of porosity on the mechanical properties of additively manufactured components: A critical review, Mater Res Express. 2020;7(12):122001.

[49] Ma CP, Guan YC, Zhou W. Laser polishing of additive manufactured Ti alloys, Opt Lasers Eng. 2017;93:171–177.

[50] Liu Y, Zhang H, Wang J, Zhao Q, Li C, Chen X et al Improving surface quality and superficial microstructure of LDED Inconel 718 superalloy processed by hybrid laser polishing, J Mater Process Technol. 2022;300:117428.

[51] Al-Maharma AY, Patil SP, Markert B. Effects of porosity on the mechanical properties of additively manufactured components: A critical review, Mater Res Express. 2020;7(12):122001.

[52] Kan WH, Ding R, Lu W, Yu Y, Li S, Liu Y. et al. A critical review on the effects of process-induced porosity on the mechanical properties of alloys fabricated by laser powder bed fusion, J Mater Sci. 2022;57(21):9818–9865.

[53] Li C, Liu Z, Fang X, Guo Y. Residual stress in metal additive manufacturing, Procedia CIRP. 2018;71:348–353.

[54] Carpenter K, Tabei A. On residual stress development, prevention, and compensation in metal additive manufacturing, Materials (Basel). 2020;13(2):255.

[55] Bastola N, Jahan MP, Rangasamy N, Rakurty CS. A review of the residual stress generation in metal additive manufacturing: Analysis of cause, measurement, effects, and prevention, Micromachines (Basel). 2023;14(7):1480.

[56] Ma E, Zhu T. Towards strength–ductility synergy through the design of heterogeneous nanostructures in metals, Mater Today. 2017;20(6):323–331.

[57] Dong Z, Ouyang P, Zhang S, Liu L, Li H, Wu Y. Effect of building direction on anisotropy of mechanical properties of GH4169 alloy fabricated by laser powder bed fusion, Mater Sci Eng A. 2023;862:144430.

[58] Bocanegra-Bernal M. Hot isostatic pressing (HIP) technology and its applications to metals and ceramics, J Mater Sci. 2004;39(21):6399–6420.

[59] Du Plessis A, Macdonald E. Hot Isostatic Pressing in metal additive manufacturing: X-ray tomography reveals details of pore closure, Addit Manuf. 2020;34:101191.

[60] Munther M, Martin T, Tajyar A, Hackel L, Beheshti A, Davami K. Laser shock peening and its effects on microstructure and properties of additively manufactured metal alloys: A review, Eng Res Express. 2020;2(2):022001.

[61] Zhang C, Dong Y, Ye C. Recent developments and novel applications of laser shock peening: A review, Adv Eng Mater. 2021;23(7):2001216.

[62] Zhou J, Zhou X, Li H, Hu J, Han X, Liu S. In-situ laser shock peening for improved surface quality and mechanical properties of laser-directed energy-deposited AlSi10Mg alloy, Addit Manuf. 2022;60:103177.

[63] Seifi M, Salem AA, Satko DP, Ackelid U, Semiatin SL, Lewandowski JJ. Effects of HIP on microstructural heterogeneity, defect distribution and mechanical properties of additively manufactured EBM Ti-48Al-2Cr-2Nb, J Alloys Compd. 2017;729(1118):35.

[64] Fu J, Li H, Song X, Fu MW. Multi-scale defects in powder-based additively manufactured metals and alloys, J Mater Sci Technol. 2022;122:165–199.

[65] Gogulraj G, Rajamurugan G.. A review on additive manufacturing of dissimilar metals: Current challenges, industrial applications, and AI/ML integration strategies, Phys Scr. 2025;[ahead of print].

[66] Romano T, Bellandi A, Nappi M, Boscolo I, Zennaro R, Michelato P. et al. Metal additive manufacturing for particle accelerator applications, Phys Rev Accel Beams. 2024;27(5):054801.

[67] Fan P, Zhong M. Laser surface micro-nano structuring via hybrid process, Handbook of Laser Micro-Nano-Engineering, 2021;937–978.

[68] Wu G, Zhang Y, Li J, Liu X, Wang H, Sun J. et al. Process and corrosion property of laser-chemical hybrid polishing of TC4 alloy, J Manuf Process. 2023;101:1374–1382.

[69] Kovalev M, Smirnov V, Dunaevsky G, Petrov V, Ivanova L, Sokolov D. et al. Efficient broadband light-trapping structures on thin-film silicon fabricated by laser, chemical and hybrid chemical/laser treatments, Materials (Basel). 2023;16(6):2350.
[70] Kumar PV, Kulkarni A, Mendhe D, Keshar DK, Babu SBT, Rajesh N. AI-optimized Hardware Design for Internet of Things (IoT) devices, In Proceedings of the 2024 5th International Conference on Recent Trends in Computer Science and Technology (ICRTCST). IEEE; 2024. p. 21–26.
[71] Habeeb HA, Wahab DA, Azman AH, Alkahari MR. Design optimization method based on artificial intelligence (hybrid method) for repair and restoration using additive manufacturing technology, Metals (Basel). 2023;13(3):490.
[72] Farshidianfar MH. Real-time Closed-loop Control of Microstructure and Geometry in Laser Materials Processing. Doctoral thesis. Waterloo (ON): University of Waterloo; 2017.
[73] Lafirenza M. Development of In-situ Optical Sensor-based Monitoring Methodologies for Additive Manufacturing Processes. Doctoral thesis. Bari: Politecnico di Bari; 2025.
[74] Van Blitterswijk RH. Adaptive Model Predictive Control for Microstructure Control in Laser Material Processing. Doctoral thesis. Waterloo (ON): University of Waterloo; 2024.
[75] Debnath S, Reddy MM, Yi QS. Environmental friendly cutting fluids and cooling techniques in machining: A review, J Clean Prod. 2014;83:33–47.
[76] Gupta K, Gupta MK. Developments in nonconventional machining for sustainable production: A state-of-the-art review, Proc Inst Mech Eng Part C J Mech Eng Sci. 2019;233(12):4213–4232.
[77] Gabriel DH, Parthiban M, Kantharaj I, Beemkumar N. A review on sustainable alternatives for conventional cutting fluid applications for improved machinability, Mach Sci Technol. 2023;27 (2):157–207.
[78] Wang B, Castellana J, Melkote SN. A hybrid post-processing method for improving the surface quality of additively manufactured metal parts, CIRP Ann. 2021;70(1):175–178.
[79] Syrlybayev D, Seisekulova A, Talamona D, Perveen A. The post-processing of additive manufactured polymeric and metallic parts, J Manuf Mater Process. 2022;6(5):116.
[80] Pourrahimi S, Hof LA. On the post-processing of complex additive manufactured metallic parts: A review, Adv Eng Mater. 2024;26(10):2301511.
[81] Jain V, Sidpara A, Balasubramaniam R, Lodha G, Dhamgaye V, Shukla R. Micromanufacturing: A review – part I, Proc Inst Mech Eng Part B J Eng Manuf. 2014;228(9):973–994.
[82] Shao Y. Investigation on the Multiscale Multiphysics based Approach to Modelling and Simulation in Abrasive Flow Machining of Aerofoil Structures and its Application Perspectives. Doctoral thesis. London: Brunel University London; 2020.
[83] Jiang Q. Material Removal Mechanism and Surface Integrity of Wrought and SLM-ed Ti6Al4V Alloys in Ultra-high-speed Machining. Doctoral thesis. Hong Kong: The Hong Kong Polytechnic University; 2024.
[84] Sose AT, Joshi SY, Kunche LK, Wang F, Deshmukh SA. A review of recent advances and applications of machine learning in Tribology, Phys Chem Chem Phys. 2023;25(6):4408–4443.
[85] Howard MJ. Development of a Machine-tooling-process Integrated Approach for Abrasive Flow Machining (AFM) of Difficult-to-machine Materials with Application to Oil and Gas Exploration Components. Doctoral thesis. London: Brunel University London; 2014.
[86] Ge J, Pillay S, Ning H. Post-process treatments for additive-manufactured metallic structures: A comprehensive review, J Mater Eng Perform. 2023;32(16):7073–7122.
[87] Sergi A. Hot Isostatic Pressing of High Temperature Materials. Doctoral thesis. Birmingham: University of Birmingham; 2022.
[88] Babamiri BB, Indeck J, Demeneghi G, Cuadra J, Hazeli K. Quantification of porosity and microstructure and their effect on quasi-static and dynamic behavior of additively manufactured Inconel 718, Addit Manuf. 2020;34:101380.

[89] Bilgin M, Karabulut Ş, Karakoç H, Kayır Y, Sarıkaya M. An experimental investigation on machining-induced surface/subsurface characteristics of Nickel-based Inconel 718 alloy: A novel hybrid approach in milling process, Tribol Int. 2024;191:109120.
[90] Yang Z, Yang M, Sisson R, Li Y, Liang J. A Machine-learning model to predict tensile properties of Ti6Al4V parts prepared by laser powder bed fusion with hot isostatic pressing, Mater Today Commun. 2022;33:104205.
[91] Yadav A, Ghazaly NM, Askar S, Alzubaidi LH, Almulla AA, Al-Tameemi AR. Machine learning-powered analysis of hot isostatic pressing for Ti-6Al-4V powder, Appl Phys A. 2024;130(9):610.
[92] Zaki S, Zhang N, Gilchrist MD. Electropolishing and shaping of micro-scale metallic features, Micromachines. 2022;13(3):468.
[93] Chen X. et al. Review on chemical mechanical polishing for atomic surfaces using advanced rare earth abrasives, J Phys D Appl Phys. 2024;58(2):023004.
[94] Abbas A, Adesina AY, Suleiman RK. Influence of organic acids and related organic compounds on corrosion behavior of stainless steel: A critical review, Metals. 2023;13(8):1479.
[95] Amjadian B. Advanced Descaling Techniques for High-temperature Oxidation in Stainless-steel Alloys. Master's thesis. Lund: Lund University; 2025.
[96] Hanzlik J. et al. The impact of surface roughness on conformal cooling channels for injection molding, Materials. 2024;17(11):2477.
[97] Martínez RZ. et al. Comparative analysis of machining and electropolishing for surface quality improvement of shape memory nitinol samples additively manufactured by laser powder bed fusion, Addit Manuf Lett. 2025;12:100261.
[98] Zhou L. et al. Research on surface integrity of Ti-6Al-4V alloy with compound treatment of laser shock peening and shot peening, Vacuum. 2022;196:110717.
[99] Yong W. et al. Influence of laser shock peening on surface integrity and tensile property of high strength low alloy steel, Chin J Aeronaut. 2021;34(6):199–208.
[100] Haribaskar R, Kumar TS. The impact of successive laser shock peening on surface integrity and residual stress distribution of laser powder-bed fused stainless steel 316L, Phys Scr. 2024;99(5):055929.
[101] Xin R, Lan L, Bai C, Gao S, He B, Wang J. Fatigue properties of selective laser melted Ti-6Al-4V alloy subjected to laser shock processing, J Mater Sci. 2022;57(21):9619–9630.
[102] Kaligar AB. et al. Femtosecond laser-based additive manufacturing: current status and perspectives, Quantum Beam Sci. 2022;6(1):5.
[103] Park JM. et al. Effect of heat treatment on microstructural heterogeneity and mechanical properties of 1% C-CoCrFeMnNi alloy fabricated by selective laser melting, Addit Manuf. 2021;47:102283.
[104] Ran X. et al. Effects of post homogeneity heat treatment processes on microstructure evolution behavior and tensile mechanical properties of laser additive manufactured ultrahigh-strength AerMet100 steel, Mater Sci Eng A. 2018;723:8–21.
[105] Heuer V, Bolton D, Friedel J, Garcia O. Advanced distortion control for case hardening of transmission components, AGMA Tech Pap, 2023;23FTM07.
[106] Aurrekoetxea M, Llanos I, Zelaieta O, López de Lacalle LN. Towards advanced prediction and control of machining distortion: A comprehensive review, Int J Adv Manuf Technol. 2022;122(7):2823–2848.
[107] Zulić S. et al. Fatigue life enhancement of additive manufactured 316L stainless steel by LSP using a DPSS laser system, Surf Eng. 2022;38(2):183–190.
[108] Zhou L, Long C, He W, Tian L, Jia W. Improvement of high-temperature fatigue performance in the Nickel-based alloy by LSP-induced surface nanocrystallization, J Alloys Compd. 2018;744:156–164.
[109] Tang J, Soua S, Mares C, Gan TH. An experimental study of acoustic emission methodology for in-service condition monitoring of wind turbine blades, Renew Energy. 2016;99:170–179.
[110] Chaghazardi Z, Wüthrich R. Electropolishing of additive manufactured metal parts, J Electrochem Soc. 2022;169(4):043510.

[111] Fayazfar H, Sharifi J, Keshavarz MK, Ansari M. An overview of surface roughness enhancement of additively manufactured metal parts: a path towards removing the post- print bottleneck for complex geometries, Int J Adv Manuf Technol. 2023;125(3):1061–1113.

[112] Conde IR. Femtosecond laser-enabled precision additive manufacturing: Process parameter optimization and surface roughness reduction, 2024.

[113] Wu G. et al. Effect of laser surface melting on superficial morphology and properties of the electrochemical polishing Ti6Al4V alloy, J Mater Res Technol. 2024;33(7434):47.

[114] Zhao C, Qu N, Tang X. Electrochemical mechanical polishing of internal holes created by selective laser melting, J Manuf Process. 2021;64:1544–1562.

[115] Guidat M. In Situ monitoring of electrochemical interfaces by reflection anisotropy spectroscopy, 2024.

[116] Haghighi A. Smart process-aware tolerance design and quality assurance for additive and hybrid manufacturing, 2020.

[117] Kantaros A, Ganetsos T, Petrescu FIT, Ungureanu LM, Munteanu IS. Post-production finishing processes utilized in 3D printing technologies, Processes. 2024;12(3):595.

[118] Hung W. Postprocessing of additively manufactured metal parts, J Mater Eng Perform. 2021;30 (9):6439–6460.

[119] Mugwagwa L. Investigation and management of residual stresses in selective laser melting of maraging steel, 2019.

[120] Chinna R, Sharma P, Fernandes F. Thermo plus mechanical based sequential machining processes, In Advancements in Machining Processes Through Hybrid and Sequential Approaches. CRC Press; p. 109–126.

[121] Manshina AA. et al. The second laser revolution in chemistry: Emerging laser technologies for precise fabrication of multifunctional nanomaterials and nanostructures, Adv Funct Mater. 2024;34 (40):2405457.

[122] Murzin SP. Artificial intelligence-driven innovations in laser processing of metallic materials, Metals. 2024;14(12):1458.

[123] Aivaliotis P, Georgoulias K, Chryssolouris G. The use of digital twin for predictive maintenance in manufacturing, Int J Comput Integr Manuf. 2019;32(11):1067–1080.

[124] Kumar KB, Rajitha A, Rao AK, Alam K, Albawi A, Sethi G. Smart materials for biomedical applications: Advancements and challenges, In E3S Web of Conferences. EDP Sciences; 2023. p. 01133.

[125] Thompson MK. et al. Design for additive manufacturing: trends, opportunities, considerations, and constraints, CIRP Ann. 2016;65(2):737–760.

[126] Lu Y. Industry 4.0: A survey on technologies, applications and open research issues, J Ind Inf Integr. 2017;6:1–10.

[127] Dávila JL, Neto PI, Noritomi PY, Coelho RT, Da Silva JVL. Hybrid manufacturing: A review of the synergy between directed energy deposition and subtractive processes, Int J Adv Manuf Technol. 2020;110(11):3377–3390.

[128] Hafeez MA, Procacci A, Coussement A, Parente A. Challenges and opportunities for the application of digital twins in hard-to-abate industries: A review, Resour Conserv Recycl. 2024;209:107796.

[129] Jafar MR, Tripathi NM, Yadav M, Nasato DS. Additive manufacturing in the age of industry 4.0 and beyond, In Advances in Pre- and Post-Additive Manufacturing Processes. CRC Press; 2024. p. 213–230.

[130] Sala D, Richert M. Perspectives of additive manufacturing in 5.0 industry, Materials. 2025;18(2):429.

[131] Rabalo MÁ, Rubio EM, Agustina B, Camacho AM. Emerging trends in hybrid additive and subtractive manufacturing. Appl Sci. 2025;15(11):6102. https://doi.org/10.3390/app15116102

[132] Javaid M, Haleem A, Singh RP, Suman R. (2023) An integrated outlook of Cyber–Physical Systems for Industry 4.0: Topical practices, architecture, and applications. Green Technologies and Sustainability, 1 (1), 100001. https://doi.org/10.1016/j.grets.2022.100001

[133] Manchili, S.K., Pezzotti, F., Dartfeldt, E. et al. Chemical and electrochemical surface modification and fatigue response of PBF-LB/M 316L stainless steel. Prog Addit Manuf 11, 2563–2577 (2026). https://doi.org/10.1007/s40964-025-01482-x

[134] Allen Love, Saeed Behseresht, Omar Alejandro Valdez Pastrana, James Sakai, and Young Ho Park (2025) Hybrid Physical-Virtual Digital Twin System for Additive ManufacturingJournal of Advanced Manufacturing Systems, 24 (1), 1–20. https://doi.org/10.1142/S0219686725500015

[135] Dehghan, S.; Sattarpanah Karganroudi, S.; Echchakoui, S.; Barka, N. The Integration of Additive Manufacturing into Industry 4.0 and Industry 5.0: A Bibliometric Analysis (Trends, Opportunities, and Challenges). Machines 2025, 13, 62. https://doi.org/10.3390/machines13010062

[136] Addepalli, P., Sawangsri, W. & Ghani, S.A.C. Advancing Orthopedic Surgical Tools: With Global Metal Dependency Trends Shifting Toward Sustainable Bioceramic Alternatives. JOM 77, 5036–5051 (2025). https://doi.org/10.1007/s11837-025-07340-z

Mohit Kumar, Tanmay Tiwari, Chandra Shekhar Rakurty

9 Sustainability Interventions in Hybrid Manufacturing Processes

Abstract: Sustainability in manufacturing demands a complete strategy that balances environmental stewardship, resource efficiency, social equity, and economic viability. Traditional subtractive manufacturing (SM) excels in precision but generates significant material waste, often 50–70% in metalworking, and consumes high energy for hard materials and fine tolerances. Sustainable SM initiatives leverage advanced computer-aided manufacturing, coolant management, and recycling to mitigate these drawbacks, yet material efficiency and complexity constraints persist. Additive manufacturing (AM) addresses waste by depositing material only where needed, supports recycled and bio-based feedstocks, and enables lightweight, high-performance components that yield life cycle energy savings, particularly in aerospace and biomedical sectors. However, AM's limitations in surface finish, dimensional accuracy, and mechanical properties often necessitate energy-intensive post-processing. Hybrid manufacturing processes (HMPs) synergistically combine AM's near-net shaping with SM's precision machining to produce complex geometries with minimal waste and high accuracy. By consolidating deposition and finishing within a single workflow, HMPs can reduce material consumption by up to 90% compared to conventional SM, extend component life through repair and refurbishment, and advance circular economy principles. This chapter offers a comparative life cycle assessment (LCA) of HMPs against subtractive routes, evaluating social, economic, and environmental impacts. The review consists of the current LCA methodologies, identifies sustainability hotspots, and explores strategies to optimize hybrid workflows. The discussed case studies illustrate real-world applications, logistical considerations, and waste management solutions, offering actionable insights for integrating HMP technologies into sustainable manufacturing practices.

Keywords: hybrid manufacturing, sustainability, life cycle assessment, additive manufacturing, subtractive manufacturing

9.1 Introduction

Hybrid manufacturing (HM) is increasingly vital today for its ability to enhance efficiency, reduce waste, and support sustainable production through the combination of multiple machines/processes [1]. Authors have given different definitions of HM. The term has been commonly used in literature to denote various HM techniques such as hybrid machines, hybrid processes, and hybrid structures/materials/functions, as de-

 | https://doi.org/10.1515/9783111704456-009

picted in Fig. 9.1 [2]. Kozak and Rajurkar [3] emphasize that hybrid machining processes should capitalize on the synergistic benefits of combined techniques while mitigating the limitations typically encountered when these processes are used independently. An alternative interpretation is that HM aims to achieve the "1 + 1 = 3" result, which means the benefits of the hybrid processes exceed the sum of distinct processes [4]. Furthermore, the International Academy for Production Engineering (Collège International pour la Recherche en Productique – CIRP) defines hybrid processes as the controlled and simultaneous interaction of multiple energy sources or process mechanisms, aimed at significantly enhancing process performance [5]. Hybrid processes should be distinguished from hybrid machines. Hybrid machines refer to a single platform that incorporates multiple manufacturing capabilities, such as milling-turning centers or integrated 3D printing and milling systems, whereas hybrid processes emphasize the functional integration and interaction of different manufacturing techniques, not merely their colocation within one machine. Besides, hybrid structures/materials/functions integrate different materials to create enhanced or entirely new performance characteristics.

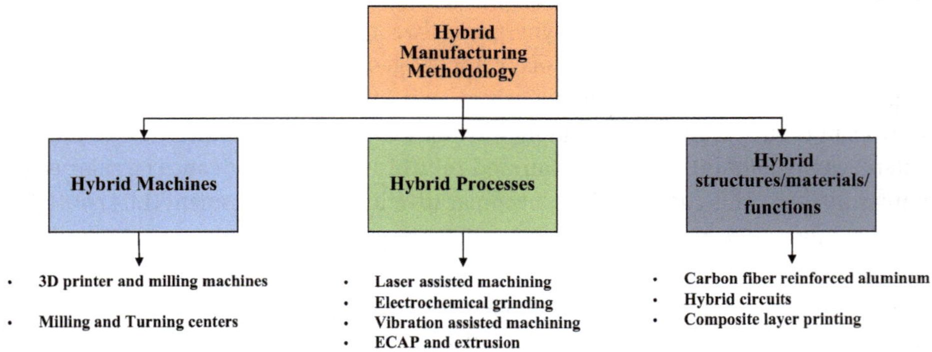

Fig. 9.1: Methodology of hybrid manufacturing.

In the last two decades, additive manufacturing (AM) has enabled the manufacturing of intricate parts that are not straightforwardly producible by conventional manufacturing methods, but its limitations are dimensional accuracy and surface quality, which constrain its applications. To overcome these limitations, a novel additive and subtractive HM approach combines the design flexibility of AM with precision machining, as depicted in Fig. 9.2 [6]. For instance, the integration of selective laser melting (SLM) with milling offers notable advantages over standalone AM and conventional machining [7]. It enables faster production through near-net-shape fabrication with precise finishing, reduces waste and tool wear, especially for rare or hard-to-machine materials, and allows the creation of complex features like internal channels. Additionally, it supports multi-material builds and enhances fatigue strength due to superior surface quality [8, 9].

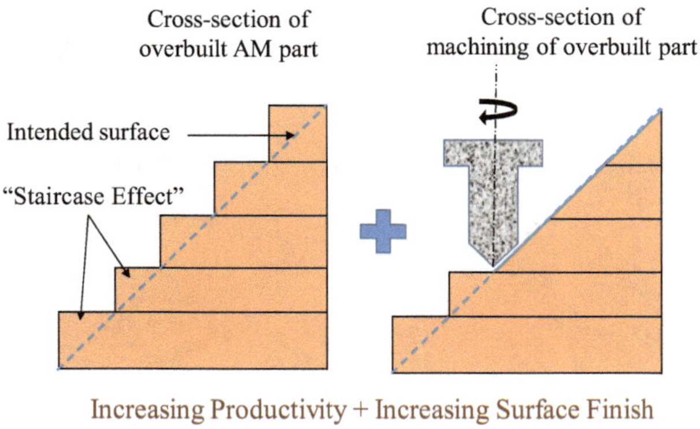

Fig. 9.2: Cross-sectional view of the machining of an overbuilt part using a combined additive and subtractive manufacturing approach [6] (with kind permission from Elsevier).

For instance, a schematic demonstration of HM (additive and subtractive) is depicted in Fig. 9.3 [6]. The HM processes (HMPs) begin with a computer-aided design model sliced into thin layers, which are sequentially built using SLM. The metal alloy powder is deposited over a build platform, and a laser selectively melts and fuses the powder layer by layer. A milling tool is used after specific layers of the deposited material to improve the surface finish of internal channels. An alternative additive and subtractive process is performed until the completion of the near-net-shaped 3D part [6]. Consequently, HMPs can produce precise and complex parts directly for use in minimum production time with enhanced mechanical properties. On the other hand, one question arises: Are HMPs sustainable? HMPs are more capable and proficient in improving part's quality, strength, and reducing material waste than the counter-process. Still, their sustainability depends on material usage, system complexity, and energy consumption. Hence, there is a requirement for careful optimization and assessment of relevant critical factors.

The term "sustainability" was first used in forestry to refer to the idea of never harvesting more than the forest can naturally grow back [10]. The term "Nachhaltigkeit," which means sustainability in German, was first used in this way in 1713 [11]. But the basic idea goes back much further; our ancestors probably worried about running out of animals to hunt, and early farmers had to think about keeping their soil healthy so they could grow food year after year. It is always been about using natural resources wisely so they don't run out. In 1987, the United Nations Brundtland Commission defined sustainability as "development that meets the needs of the present without compromising the ability of future generations to meet their own needs" [12]. Whereas environmental issues are essential, the main argument revolves around wel-

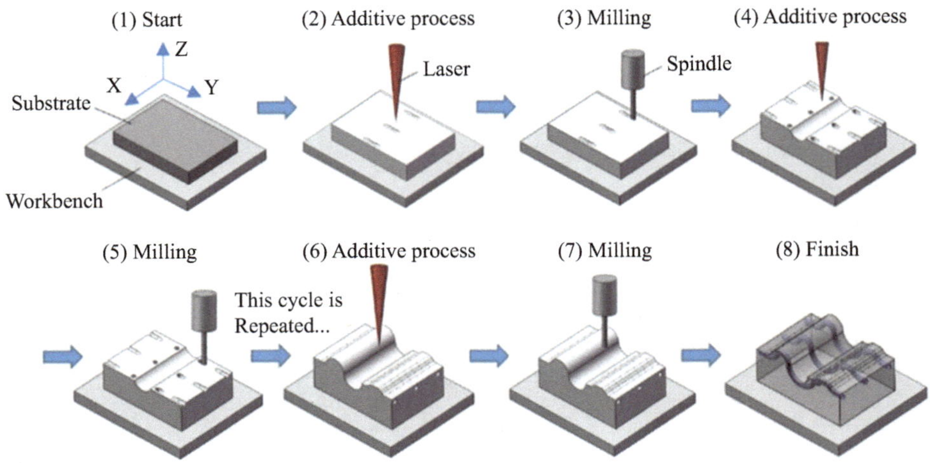

Fig. 9.3: Schematic representation of hybrid (additive and subtractive) manufacturing [6] (with kind permission from Elsevier).

fare and intergenerational justice. We should protect the environment not for its intrinsic value, but to ensure that resources are available for future generations. A key development in sustainability is its interpretation through three interconnected dimensions such as social, economic, and environmental, which must function in harmony.

Sustainability in manufacturing processes is becoming a growing priority for present-day manufacturing organizations due to various developed and emerging factors, including concerns about the environment, higher energy expenses, stringent regulations, depletion of natural resources, and growing consumer demand for products that are environmentally friendly [13, 14]. Therefore, manufacturing companies worldwide are interested in low carbon emissions or environmentally friendly production. Currently, the emphasis is on production, which differs from conventional manufacturing to life cycle assessment (LCA) and pollution control. Processes and products inevitably interact with the environment, causing harmful impacts such as pollution and resource damage [15]. In many developed countries, a considerable amount of industrial waste is not recycled or managed appropriately, contributing to serious problems such as hazardous waste accumulation, difficulties locating new landfill sites, and rapid depletion of natural resources. This unsustainable approach emphasizes the urgent need for more effective trash management. Robert U. Ayres proposes a constructive alternative in which firms take a circular strategy, converting trash from one area into usable input for another [16]. This not only reduces environmental effects but also encourages resource efficiency and cross-industry collaboration, opening the door for more resilient industrial ecosystems. In this context, a study applying a hierarchic metric approach to integrate environmental considerations into manufacturing, specifically for paper recycling, emphasizes the importance of a product life cycle perspective. It outlines key strategies such as environmentally conscious products, material selection, recycling, efficient product design and manufacturing, robust waste collection methods, effective scrap recovery, and the safe disposal of harmful materials and components [15]. Thus, sustainability in manufacturing has become a critical necessity for both existing and forthcoming generations. Recognizing this necessity, the factory of the future is evolving beyond a predominantly economic focus to embrace a more holistic approach, addressing environmental and social challenges alongside economic goals and integrating all three pillars of sustainability into its core manufacturing strategies [17].

The economic pillar of sustainability emphasizes minimizing overall system costs and optimizing factors that directly affect economic performance. The environmental dimension, often evaluated through LCA, examines the negative impacts of a system, including contributions to global warming, toxicity levels, and depletion of natural resources. Meanwhile, the social aspect focuses on ensuring worker health and safety, promoting employee well-being, and meeting customer satisfaction, thereby supporting a balanced and responsible manufacturing approach [18]. In addition, sustainable manufacturing encompasses all four stages of the product life cycle: pre-manufacturing stage, manufacturing, usage, and post-use and follows the 6R principles such as reuse, reduce, recycle, recover, redesign, and remanufacture [19]. To achieve sustainability in production, the primary components (product, process, and system) must not only improve energy and resource efficiency and decrease environmental impact but also ensure operational safety, promote human well-being, and maintain or improve product and process quality [19].

The preceding discussion explained the importance of HMPs in detail and their alignment with sustainability goals. The following section will delve into sustainability assessment frameworks alongside relevant studies and case analyses that demonstrate sustainability interventions in HMPs.

9.2 Sustainability Assessment Framework

This section employs a mixed-method approach, integrating quantitative and qualitative techniques, to develop a comprehensive sustainability assessment framework in HM. Sustainability is broadly categorized into three dimensions: economic, social, and environmental. The economic dimension focuses on cost optimization and other financially relevant indicators. The environmental evaluation, typically conducted via LCA, examines impacts such as global warming potential (GWP), toxicity, and resource depletion [20–22]. The social dimension considers worker health and safety, as well as stakeholder satisfaction. For social assessment, a targeted questionnaire may be distributed among HM professionals to collect relevant data. The collected data can then be analyzed using the entropy weight method (EWM) to normalize indicators, followed by the assignment of weights to various indicators and subsystems to compute an overall sustainability index. To ensure the robustness of the framework, sensitivity analysis can be performed by varying normalization methods. This structured methodology aids decision-makers in selecting sustainable manufacturing routes based on the derived sustainability index. The framework development follows three key steps: evaluation of economic, environmental (LCA), and social factors for HM and SM (subtractive manufacturing) routes; (2) indicator selection and sustainability index calculation via the EWM; and (3) validation through sensitivity analysis [18].

System boundary selection: The sustainability assessment for HM and SM can adopt different system boundaries, such as gate-to-gate, gate-to-grave, or cradle-to-grave, depending on the scope of analysis. As illustrated in Fig. 9.4, the system boundary encompasses all relevant inputs (energy and materials) and outputs (final product), including stages like transportation, usage, and end of life. While transportation and use-phase impacts may be similar for both manufacturing routes, the material processing and manufacturing stages differ significantly. A carefully selected functional unit, representing a product manufactured by both HM and SM, is crucial for ensuring meaningful comparisons across the life cycle stages.

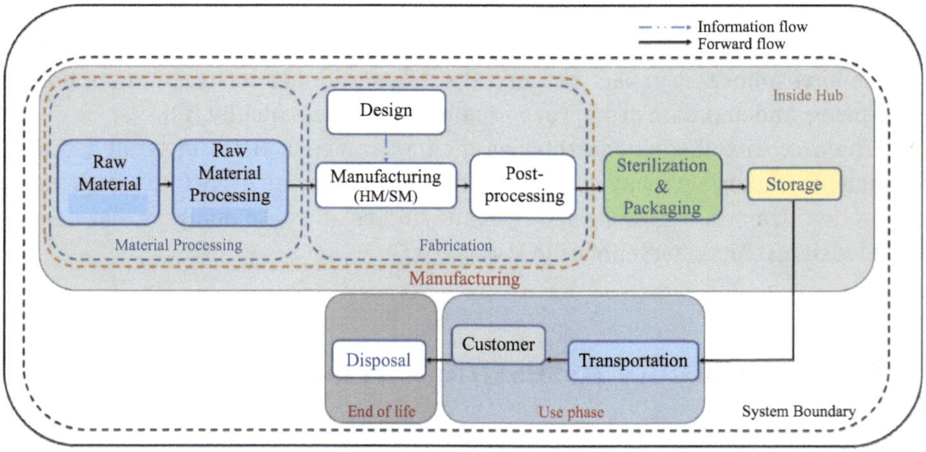

Fig. 9.4: Life cycle stages of products manufactured through HM/SM [18] (with kind permission from Elsevier).

9.2.1 Economic Analysis

Economic sustainability is a key pillar in evaluating the overall sustainability of a manufacturing process, as the adoption of any production method often depends on its economic feasibility [23]. This involves assessing all associated costs across the product's life cycle, including raw materials, electricity, labor, equipment depreciation, manufacturing, transportation, storage, and disposal. Additionally, indirect factors such as production rate, energy consumption, and capacity utilization are essential to understand the true economic impact. For instance, in gear manufacturing, the unit product cost captures expenses across stages such as material processing, fabrication, post-processing, packaging, and user-phase transportation, with disposal focused on packaging waste and recycling or repairing. The total cost for a manufacturing route can be evaluated by combining all the detailed cost components. A decision can be taken by comparing the total cost involved in two or more manufacturing routes along with environmental sustainability.

9.2.2 Environmental Analysis

Environmental impact analysis of a product or all its manufacturing stages can be assessed using LCA software tool. Several tools are commercially available for conducting LCA, including SimaPro, GaBi Sphera, OpenLCA, OneClick, and Ecochain. LCA helps map the environmental footprint across all stages of a product's life cycle from raw material extraction to end-of-life disposal, which reveals key areas where sustainability can be improved [20]. According to the standardized method recognized under ISO 14040/14044, the process involves four steps: describing the goal and scope, gathering life cycle inventory (LCI) data, piloting impact assessment (IA), and interpreting the results, as depicted in Fig. 9.5. These structured steps can be utilized by HM and SM processes to analyze the environmental impact individually or comparatively.

9.2.2.1 Goal and Scope Definition

This stage plays a vital role in establishing the foundation for a comprehensive LCA. In this context, the system boundary encompasses all major stages of the product's life cycle, including raw material acquisition and processing, manufacturing operations, packaging, storage, transportation, and end-of-life scenarios such as waste generation, recycling, and disposal. To ensure an accurate and meaningful comparison between HM and SM, a cradle-to-grave assessment methodology is recommended [24, 25]. This approach provides all the environmental affiliations of a product from raw material mining to the final disposal. However, a system boundary for LCA should be selected according to our scope of work. Apart from system boundary, a proper selection of functional unit is also important for evaluating and comparing environmental impact for different manufacturing route. Here the main objective of goal and scope of the study are aligning functional unit and system boundary to identify the eco-friendly and resource-efficient manufacturing strategies.

9.2.2.2 Life Cycle Inventory Analysis

LCI is a backbone of LCA of any product or process, which requires quantitative data of input/output materials and energy resources for the defined boundary system [18]. In-depth inventory analysis requires energy consumption, raw material use, water or chemical use, waste generation, and environmental emissions, such as hazardous element emissions to water, soil, and air, throughout the life cycle of a product. These inventory data can be obtained directly from documentation, measurement, and observation during actual production in a laboratory or industry, which involves electricity consumption, machine operation time, raw materials used, waste material throughout the production, etc. In case, data is difficult to obtain or unavailable, then secondary data can be utilized from the commercially available LCA tool from its database library. The acquired inventory data should be accurate, because it directly affects the reliability of environmental assessment results. Therefore, inventory data should be close to real-world applications as possible and for transparency, assumptions, approximation, and data gaps must be included in documents for transparency.

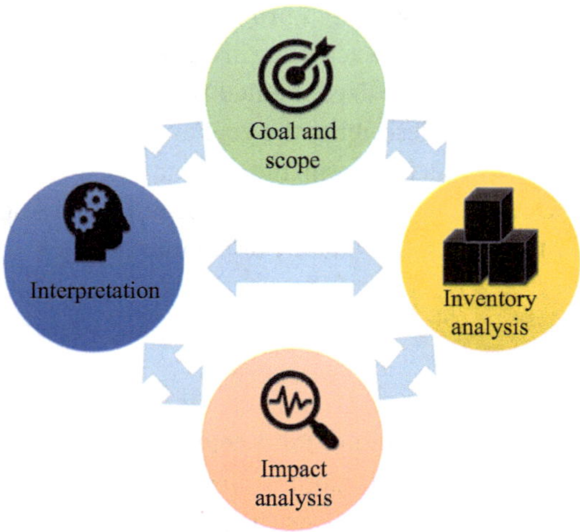

Fig. 9.5: Four phases of LCA according to ISO 14040/14044 [18] (with kind permission from Elsevier).

9.2.2.3 Impact Assessment

The IA phase converts the inventory data into quantitative and qualitative environmental IA of a product or system. This step builds upon the LCI by categorizing the various outputs, such as emissions and resource use into environmental impact categories, each represented by specific indicators.

Using LCA tools such as SimaPro, the modeled system can be evaluated using established global IA methodologies. Some of the widely applied methods available in SimaPro include IMPACT World+ (Endpoint and Midpoint), LC-IMPACT (with multiple versions), and ReCiPe 2016 (both Endpoint and Midpoint approaches across three perspectives: Individualist, Hierarchist, and Egalitarian).

For instance, the "ReCiPe 2016 Midpoint (H)" method is employed to assess the environmental load. The "H" stands for the Hierarchist perspective, which serves as the default option due to its balanced and policy-relevant approach, representing a scientific consensus on impact modeling and time horizon. ReCiPe 2016 Midpoint (H) encompasses 18 midpoint impact categories that collectively address effects on ecosystems, human health, and resource depletion. These categories include global warming, stratospheric ozone depletion, ozone formation, ozone formation human health, terrestrial ecosystems, ionizing radiation, fine particulate matter, human carcinogenic toxicity, marine ecotoxicity, fossil resource scarcity, human non-carcinogenic toxicity, mineral resource scarcity, terrestrial acidification, freshwater eutrophication, terrestrial ecotoxicity, freshwater ecotoxicity, marine eutrophication, land use, and water consumption [20].

These indicators provide a comprehensive view of the environmental implications across air, water, land, and biological systems. The IA phase, therefore, plays a critical role in identifying key environmental hotspots and supporting informed decisions toward more sustainable product and process design.

9.2.2.4 Interpretation

Interpretation is the process of identifying, evaluating, and verifying the results of the LCI and IA. The steps for life cycle interpretation are as follows: (i) identify critical issues; (ii) evaluate the data's completeness, sensitivity, and consistency; and (iii) draw conclusions and provide recommendations [25]. The significant issues are the data elements that have the greatest impact on the LCI and LCI analysis results. These can include specific inventory parameters, impact categories, or life cycle stages like manufacturing. The second stage includes checks for completeness, sensitivity of major data items, and consistency. Such checks are essential for consistent and accurate results. In most circumstances, there is no obvious answer to the question of which product option is best for the environment. This could be due to uncertainty in the results or different outcomes in the impact categories that were considered meaningful. The LCA process can still provide important results in terms of gaining a better knowledge of the environmental implications of each product alternative (e.g., which impacts occur where).

9.2.3 Social Analysis

Social analysis is a third pillar of sustainability, which is used to assess the social impacts of a product or process across the life cycle of a product or complete process. It examines how production activities affect human workers, customers, and all related community. Key focus areas included worker health and safety, and job satisfaction for both employees and customers:

a) **Worker health and safety** involves safeguarding employees' physical, mental, and social well-being by preventing occupational injuries and illnesses.
b) **Job satisfaction** covers aspects such as working hours, fair wages, equal opportunities, job security, and access to training and education.
c) **Customer satisfaction** pertains to product safety, user privacy, and protection from potential risks related to the product or process.

To evaluate these social aspects, qualitative indicators can be quantified using a Likert scale from 1 (negative impact) to 5 (positive impact). Data can be gathered through expert surveys, involving experienced professionals from industry and academia familiar with HM, AM and SM in relevant applications. A structured questionnaire can be used to capture their insights on the social impacts of a product production technology.

9.2.4 Quantitative Sustainability Assessment

The quantitative sustainability assessment combined all the data from environmental, economic, and social analyses. Each pillar comprises different indicators relevant to life cycle of a product or processes. Sustainability can be evaluated by combining all the involved indicators inside domains, and indicators' importance can be determined by various methods. One of them is EWM, which confirms an objective calculation of the sustainability index [26, 27].

EWM is used to objectively assign weights to indicators based on the variability of their data. In this regard, entropy measures the degree of uncertainty, meaning lower entropy signifies more relevant information and thus a higher weight for that indicator. EWM reduces subjective bias by relying specially on data distribution, making it well-suited for sustainability assessments that combine all indicators across economic, social, and environmental pillars. EWM has been commonly used in sustainability index evaluations due to its reliability and objectivity.

Moreover, the reliability and robustness of the developed sustainable assessment model can be enhanced by sensitivity analysis. This analysis examines how the output results are sensitive and when certain parameters change. The changes in input parameters can be evaluated by the effect on output parameters. Hence, critical variables and potential uncertainties in the system can be identified by changing statistical adjustments, decision-making methods, etc.

9.3 Sustainability Interventions in Hybrid Manufacturing

9.3.1 Energy Analysis Framework for Hybrid Versus Subtractive Manufacturing

A comparative life cycle energy demand of HM (AM + SM) and SM can be proposed as a decision-making method to determine the greatest energy-efficient method for manufacturing specific parts by considering influencing factors. This method consists of the following steps [28] and is represented in Fig. 9.6.

Step 1: The first step needs to identify part attributes, possibilities for design optimization, machinability, and use scenarios. The part attributes are important for identifying whether the part can be produced using both HM and SM processes.

Step 2: The second step is to find key factors that affect energy demand in LCA, including lightweight ratio, process parameters, and solid-to-envelope ratio (SER). For example, weight plays a critical role in aerospace applications, making the lightweight ratio a conclusive factor.

Step 3: Define the study scope and assumptions, choose among gate-to-gate (manufacturing phase), cradle-to-grave (covering raw material fabrication, manufacturing, use, and disposal or end-of-life phase), or cradle-to-gate (material fabrication and manufacturing phase only). This step also involves assumptions such as the inclusion of recycling, differences in service life, and batch production effects.

Step 4: Conduct LCI analysis by simplifying processes, creating models, and gathering data for energy consumption in each stage.

Step 5: Differentiate the energy demand of both HM and SM based on the collected data. This comparison helps identify the more energy-efficient route, recognize the most energy-intensive processes, and develop effective energy-saving strategies. It also enables identification of the break-even point, where HM becomes more advantageous than SM and guides optimization efforts to reduce total energy consumption.

This procedure provides a practical and quantitative tool for choosing the most sustainable manufacturing method tailored to specific part requirements [28].

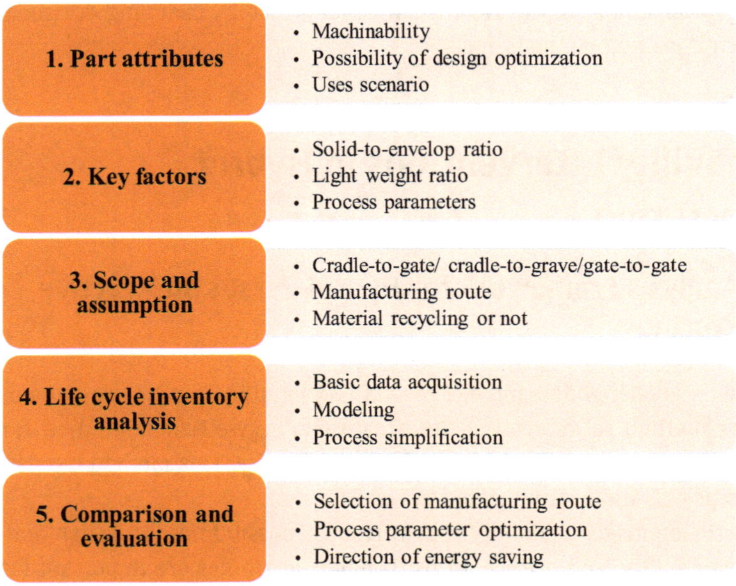

Fig. 9.6: A decision-making model for assessing energy consumption in hybrid and subtractive manufacturing.

9.3.2 Critical Key Factors for Energy-Efficient Process Selection

It is difficult to compare the energy demands of HM and traditional SM since it depends on a number of interconnected elements. Three factors, the energy reduction coefficient (ERC), lightweight ratio, and SER, significantly impact the manufacturing route's total energy efficiency. These elements affect several phases of the product's life cycle, such as the use of materials during production, weight-related advantages during usage, and the possibility of long-term energy savings. To choose the most sustainable and energy-efficient technique for a particular part, it is essential to comprehend the implications of each of these aspects alone and in combination. Below is a thorough explanation of each important component.

9.3.2.1 Solid-to-Envelope Ratio

As demonstrated in Fig. 9.7, the final part in the conventional SM process is usually machined from a billet that fits the part's bounding envelope. Particularly for parts with thin walls or intricate geometry, a significant amount of the original material is removed and discarded. The SER, a dimensionless metric developed by Watson and Taminger [29], is used to measure this material reduction. The mass of the actual portion in relation to the mass of the bounding envelope is reflected in this ratio. While a greater ratio indicates better material utilization, a lower ratio indicates more material waste during machining. In additive-subtractive HM, material is added layer by layer to approach the final shape, considerably minimizing waste, shown in Fig. 9.8. The material intake is mostly proportional to the solid fraction of the component, with minimal surplus energy required to eliminate the remaining material. As a result, there must be a break-even threshold for SER below which the HM is the most energy efficient and above which the SM is the most energy efficient.

9.3.2.2 Lightweight Ratio

One of HM's significant advantages is its ability to generate lightweight structures through topology optimization and part consolidation, allowing for substantial mass reduction while maintaining mechanical performance. The lightweight ratio is the mass of an HM-produced part in comparison to its SM counterpart. For example, a lightweight ratio of 0.17 signifies an 83% weight reduction in an engine bracket [30]. This reduction increases material efficiency, reduces manufacturing energy usage, and has substantial consequences for energy consumption during the use phase, particularly in weight-sensitive applications such as aerospace.

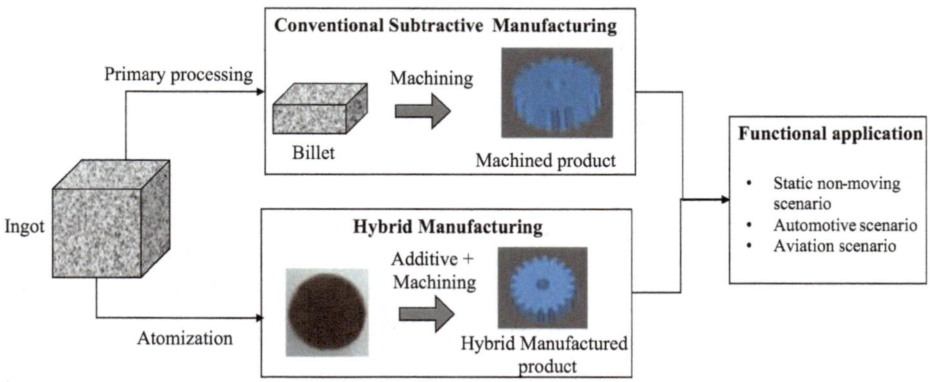

Fig. 9.7: Workflow comparison between conventional subtractive and hybrid manufacturing processes.

9.3.2.3 Use-Phase Energy Savings and Energy Reduction Coefficient (ERC)

Lightweighting influences both the production and usage phases of a product's life cycle. In transport industries like aviation and shipping, lighter components result in significant operational energy savings. For example, studies suggest that lightweight aeroplane components might save 70–173 million GJ/year in primary energy by 2050, totaling up to 2.8 billion GJ [31]. The ERC is used to characterize the sensitivity of these reductions across many applications. This variable indicates that the energy saved per unit weight decreases during the product's lifetime. The impact varies by usage scenario: airplanes and high-speed ferries save up to 100 times more energy per unit weight reduction than road vehicles due to longer working lifespans and higher energy needs.

The above description clearly shows that the lightweight ratio, ERC, and SER are highly important variables in deciding which manufacturing route (HM or SM) is more energy-efficient for a particular part. These characteristics influence not just the energy required during material processing and part manufacture, but also throughout the product's lifespan. A strong decision-making model must take into account the synergistic impacts of these elements in order to give a thorough assessment of energy efficiency and advise on the selection of the most sustainable manufacturing method.

9.3.3 Comparison Framework of HM and SM

A comprehensive framework is proposed to evaluate the crucial energy demand of intricate components manufactured via the HM additive-subtractive route or the conventional SM route, as depicted in Fig. 9.8 [19]. The HM route incorporates direct metal laser deposition with five-axis controlled computer numerical control (CNC) milling, while the SM route relies solely on subtractive techniques. The framework adopts a system boundary of cradle-to-grave encompassing material fabrication, manufacturing, use, end-of-life recycling, and transportation energy can be excluded due to its relatively minor contribution.

Assumptions play a vital role in sustainability assessment as they help define system boundaries, simplify complex processes, and address data gaps. They ensure consistency and comparability among various production paths or situations, particularly when real-world data is insufficient or unavailable. Assumptions also allow for modeling realistic consumption and end-of-life situations, directly affecting the accuracy of the demand for energy and the environmental effect. Clearly stated assumptions improve the reliability, transparency, and application of LCA results. For instance, several assumptions were made: transportation energy was neglected; recycled materials were used only to yield secondary raw materials; no service life dissimilarity between HM and SM parts; batch size effects were ignored; and both routes used the same material. The HM route involved atomizing raw materials into powder, building parts additively via laser deposition, and using milling to ensure dimensional accuracy. In contrast, the SM route started with pre-formed billets or slabs, which were machined and assembled into final components. The manufacturing phase in both routes was energy-intensive, but differed in material flow and scrap generation.

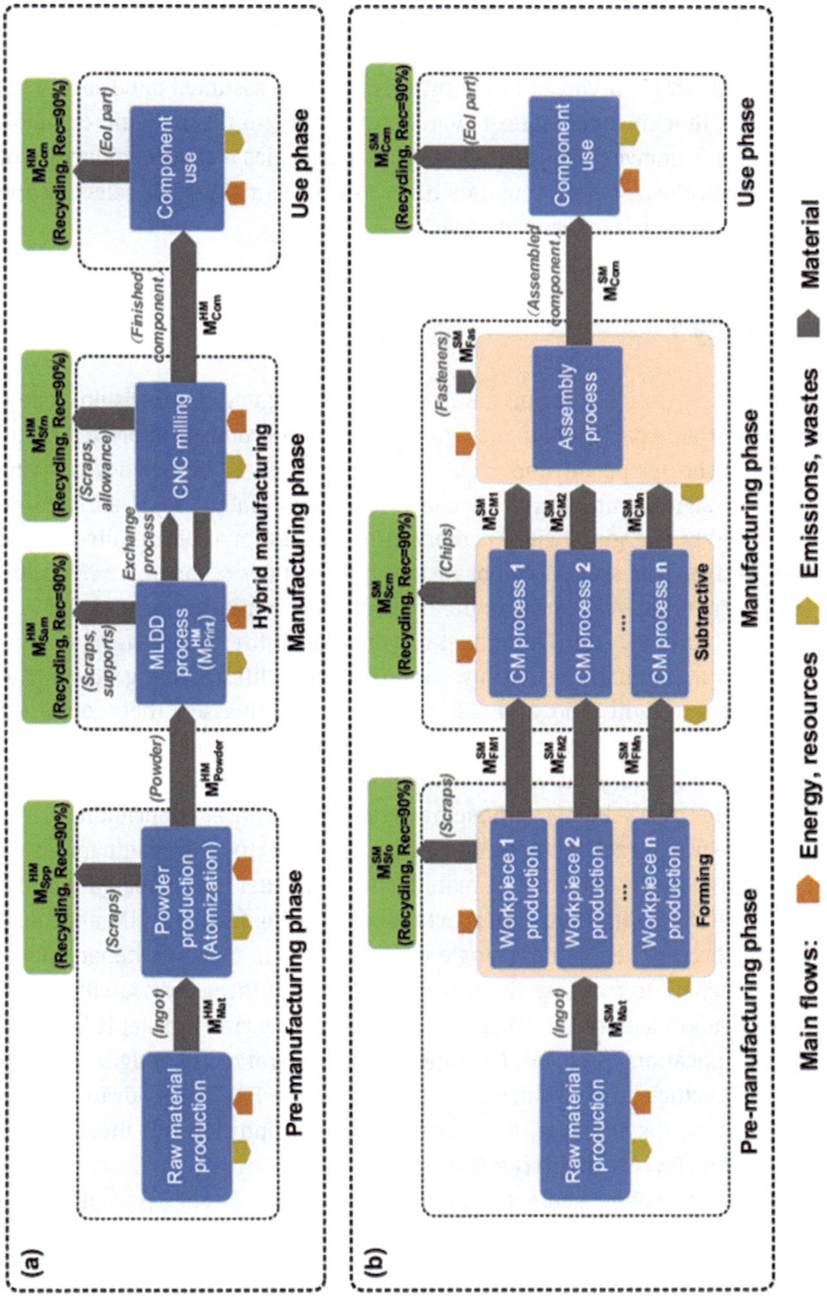

Fig. 9.8: System boundaries and energy flows of (a) HM process and (b) SM process [28] (with kind permission from Elsevier).

During the use or functional phase, energy consumption depends on the application, ranging from high fuel use in aerospace to negligible use in static components. End-of-life recycling, which involved remelting scrap, can be assumed to be consistent for both processes, though some material loss occurred due to thermal and contamination effects. This framework, illustrated in Fig. 9.8, provides a clear comparison of energy and material flows, supporting data-driven decision-making for selecting the more energy-efficient manufacturing approach.

9.3.4 Case Study 1

HM offers a promising prospect for manufacturing lightweight, high-precision, ready-to-use, or near-net-shape parts within a single setup. It also promises improved material efficiency and shorter production cycles. However, its energy efficiency remains uncertain due to significant electricity use during manufacturing. Determining whether HM is highly energy efficient compared to conventional SM requires a full life cycle perspective. This section elaborates on a case study regarding a decision-making model to quantitatively compare the life cycle energy requirement of HM and SM developed by Liu et al. [28]. The model considered all life cycle stages: (i) pre-manufacturing, (ii) manufacturing, (iii) use, and (iv) end-of-life recycling. Key variables such as ERC, lightweight ratio, and SER were integrated to assess their combined impact on energy demand. Specific energy consumption during manufacturing was modeled as a process parameter, with a defined relationship between laser power, process rate, and energy use in HM, enhancing the model's general applicability.

A simplified bearing bracket has been selected as a case study for comparison investigation, as shown in Fig. 9.9. The HM-manufactured bracket is anticipated to function similarly in service to its SM counterpart while weighing less. The HM approach combines three subcomponents into a single unit, resulting in three exchange operations during production. In contrast, the SM route consists of three individually made subcomponents that are assembled using 21 fasteners. The bearing bracket is assessed in three typical application scenarios: (1) stationary nonoperational configuration, (2) use in automotive vehicle, and (3) aircraft application. The ERC varies dramatically across these contexts; for example, it is zero in a static condition but increases to 1.125×10^5 MJ/kg over 25 years in aircraft [31].

Figure 9.10 illustrates how geometry, material strategy, and usage type affect energy performance for different scenarios. The energy demand comparison between HM and SM routes varies across usage scenarios:

(a) Static nonmoving scenario: With zero ERC, energy efficiency depends mainly on material usage and recycling. SM is more efficient, especially with recycling, but HM can be favorable at low SER (0.1–0.2) without recycling.

(b) Automotive scenario: With low ERC, lightweighting offers modest benefits. SM remains more efficient when recycling is included, while HM shows advantages only with low SER and no recycling.
(c) Aviation scenario: High ERC makes use-phase energy savings dominant. HM is clearly more efficient due to significant lightweighting benefits, regardless of recycling in SM.

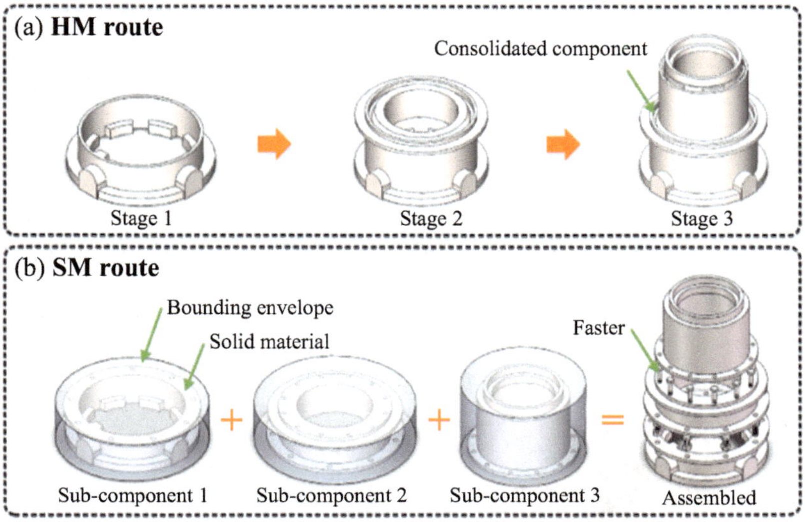

Fig. 9.9: Process schematics of bearing bracket production using HM and SM process [28] (with kind permission from Elsevier).

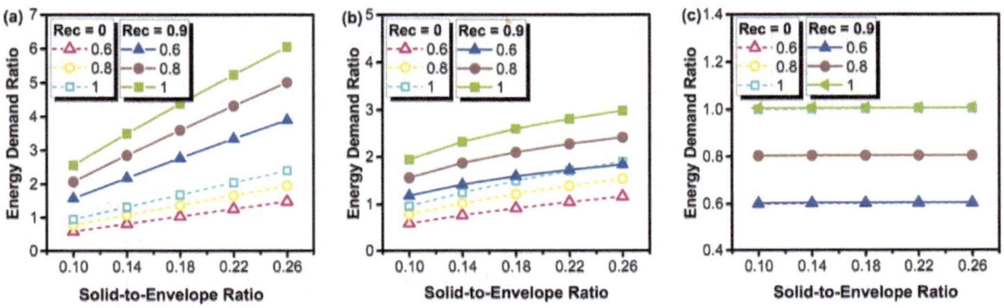

Fig. 9.10: Effect of application scenario on energy demand ratio of HM to SM routes: (a) static, (b) automotive, and (c) aviation [28] (with kind permission from Elsevier).

To better illustrate the energy requirement distribution across each phase, Fig. 9.11(a)–(d) presents the results for the static nonmoving scenario. In the HM route, the majority of energy is consumed during the metal laser deposition method in the production phase, reaching approximately 3,500 MJ. However, the SM route's highest energy consumption, approximately 3,300 MJ occurred during raw material fabrication in the phase of preproduction. The metal laser deposition stage accounts for roughly 84.5% of HM's total energy use, whereas raw material production contributes around 85.5% for SM. These findings highlight that reducing energy in SM relies heavily on material recycling, while energy savings in HM can be best achieved by optimizing metal laser deposition or AM process parameters to lower electrical consumption.

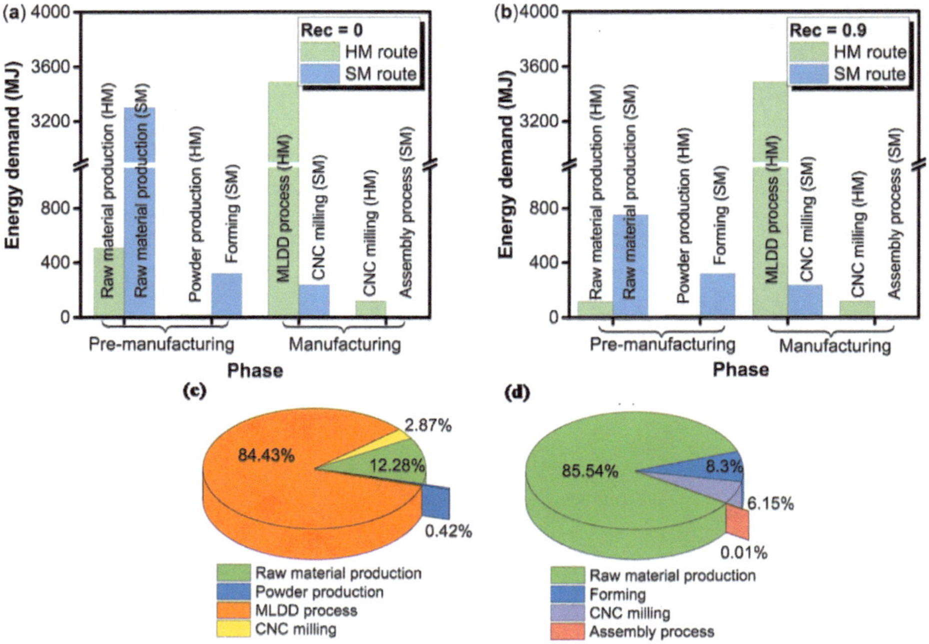

Fig. 9.11: Energy demand results for HM and SM routes in the static nonmoving scenario: (a) without material reuse and (b) with material recycling; energy distribution across life cycle phases for (c) HM route and (d) SM route under conditions of $R_{se} = 0.14$, $R_{lw} = 0.8$, and $R_{ec} = 0$ [28] (with kind permission from Elsevier).

This case study determined the decision to use life cycle-based energy comparisons of HM (additive-subtractive) and SM during the product life cycle, including preproduction, production, and use phases. The model uses an "energy demand ratio" based on three key factors: SER, ERC, and lightweight ratio. HM is particularly beneficial for aerospace components due to their low weight and high ERC, which significantly lowers energy use during the use phase. In static applications without recycling, HM's energy use is mainly from manufacturing (~84%), while SM's energy use is largely from material production (~86%). The breakeven SER ratio is 0.12 (without recycling), below which HM is more energy efficient. Increased HM's process rate from 1.5 to 5.5 g/min and reduced manufacturing energy use by 38% [28].

9.3.5 Case Study 2

This case study presents a comparative environmental sustainability evaluation of turbine blade production using directed energy deposition (DED)-based hybrid AM (HAM) and conventional CNC milling [22]. The main objective is to identify the critical contributors to environmental impact during each manufacturing process and reduce the use of those critical contributors to enhance sustainability performance. The system boundary condition was considered cradle-to-gate for LCA approach aiming to measure energy and material consumption and associated environmental implications across both manufacturing processes [32].

The system boundary comprises all main life cycle stages, including energy input, material input and output, emissions, and waste/scrap material flow from both HAM and CNC machining processes. Both manufacturing processes started with iron ore and produced a turbine blade for functional application but followed distinct production routes, as shown in Fig. 9.12(a) and (b). The HAM route involved raw material and substrate production, powder production-gas atomization, and transportation, followed by DED-based blade fabrication, heat treatment, and CNC finishing. Conversely, the CNC process includes material extraction, transportation to casting/forging facilities, steel production, delivery to the CNC lab, and blade shaping via CNC, followed by similar post-processing. The distribution, use, and end-of-life phase of two production methods are deemed equal and omitted from the scope of the boundary.

Energy consumption for part fabrication was calculated using time-series data recorded by the logger. Figure 9.13 illustrates the comparative material and energy usage for both HAM and CNC processes. The buy-to-fly ratio (BFR) is defined as the mass of the input material to the mass of the finished component, and it is significantly lower for HAM (1.34:1) compared to CNC machining (5.53:1), indicating more efficient material utilization in HAM. For energy analysis, the unit process "Electricity, lignite coal, at power plant" from the USLCI database was employed.

Environmental IA results indicated that the ecological load unit in HAM is lower than CNC for considering turbine blade production. If supposing equal product features, HAM reduces life cycle environmental impacts by 53% compared to CNC machining. As shown in Fig. 9.14(a), both methods are dominated by abiotic depletion potential (ADP) and acidification potential (AP), followed by GWP and photochemical ozone creation potential (POCP), with eutrophication potential (EP) and ozone depletion potential (ODP) contributing minimally. Figure 9.14(b) reveals that HAM's AP, POCP, EP, GWP, ADP, and ODP values are only 34.6%, 25.6%, 44.7%, 32.2%, 24.7%, and 27.2%, respectively, of those for CNC, reflecting CNC's higher energy and material waste driven by its 5.53:1 BFR versus HAM's 1.34:1. Figure 9.14(c) illustrates that electricity for HAM predominantly derives from hard coal, with smaller shares of crude oil and natural gas, owing to greater on-site energy use and minimal transpor-

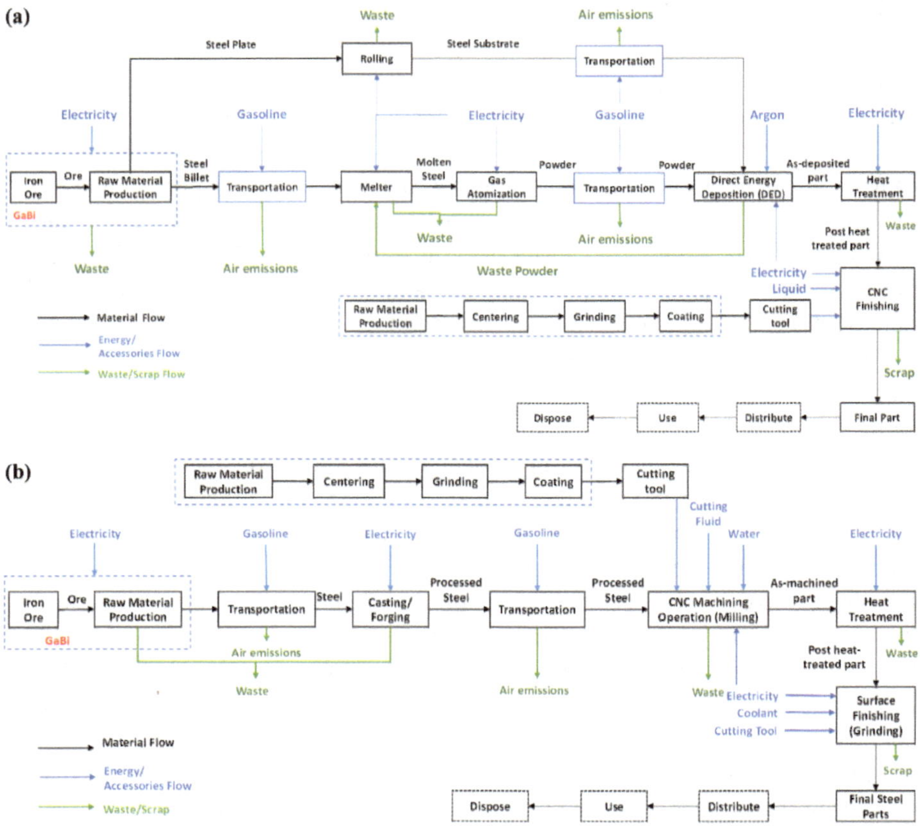

Fig. 9.12: Schematic representation of detailed work flow of (a) HAM and (b) CNC machining [32]. From *Green Manufacturing Open* (OAE Publishing Inc.), used under Creative Commons Attribution (CC BY) 4.0 License.

tation demands. In contrast, CNC's energy mix is similarly coal and oil-heavy but includes more natural gas, as its larger material transport requirements amplify resource consumption.

9.4 Conclusion and Future Directions

HM combines additive and subtractive technologies to produce complex, high-precision products, while significantly reducing material discarded and energy consumption. This chapter investigated the sustainability assessment framework for HM, including the environmental, economic, and social dimensions. Energy analysis frameworks for HM and SM highlighted critical factors that govern process efficiency, such as SER, lightweight ratio, use-phase energy savings, and energy reduction coeffi-

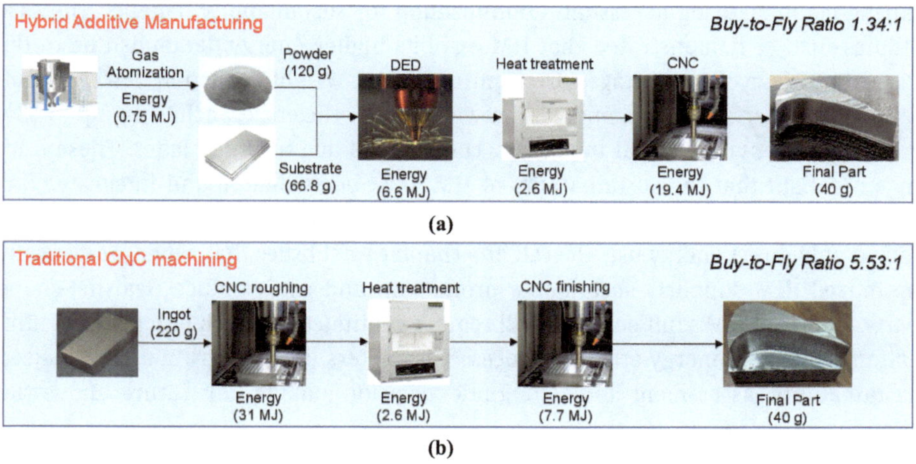

Fig. 9.13: Material and energy utilization across stages of the (a) hybrid additive manufacturing and (b) CNC machining [32]. From *Green Manufacturing Open* (OAE Publishing Inc.), used under Creative Commons Attribution (CC BY) 4.0 License.

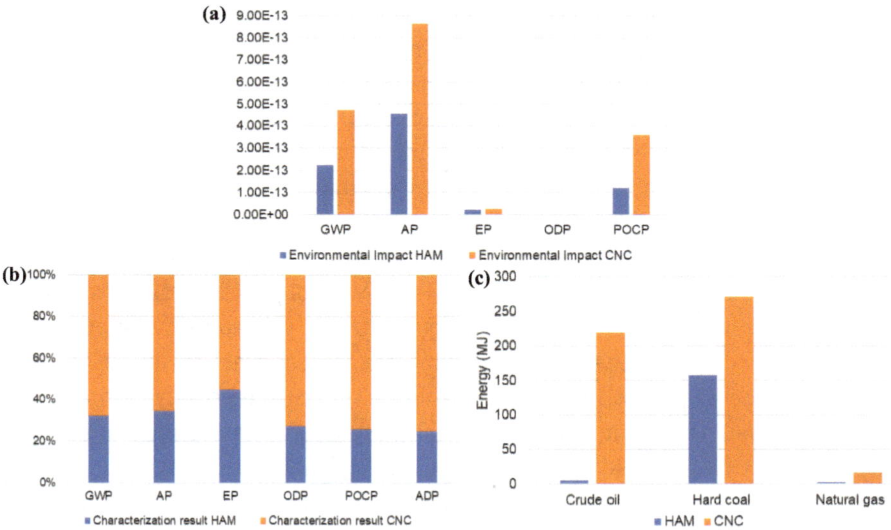

Fig. 9.14: Environmental assessment of HAM versus CNC machining: (a) normalized impact scores, (b) contribution of each impact category to the total, and (c) comparison of fossil energy consumption [32]. From *Green Manufacturing Open* (OAE Publishing Inc.), used under Creative Commons Attribution (CC BY) 4.0 License.

cients. Comparative analyses revealed HM's capability to deliver lightweight, high-precision, and material-efficient components with shorter production cycles. This can particularly benefit aerospace and automotive sectors, though its energy intensity during manufacturing necessitates optimization for sustainability benefits. The case studies further demonstrated that HM exhibits higher energy demands due to the laser. But life cycle advantages are significant, e.g., weight reduction, fewer assemblies, and material savings, and can outweigh SM's efficiencies, achieving up to 53% reduction in environmental impact for components like turbine blades. These findings highlight that the sustainability of HM must be evaluated and integrated, accounting for premanufacturing, manufacturing, use, and end of life rather than focusing on only direct energy use. Overall, the chapter establishes that, when strategically optimized, HM supports sustainable production and can produce near-net-shape parts. However, HM's full potential will require continued research into material utilization strategies, energy-efficient process parameters, and integration of advanced sustainability assessment tools to guide decision-making for future industrial adoption.

HM AM-SM aligns closely with Industry 4.0 goals enhancing flexibility, quality, and productivity through large-scale customization by combining DED with CNC milling in a single workstation. However, realizing its full potential requires overcoming several technical challenges. First, the integration of additive and subtractive processes demands efficient removal of residual powders and cutting fluids from the build platform to prevent fire hazards, abrasive wear on machine components, and loss of accuracy. Dedicated powder collection and recycling systems are essential since powder reuse rates are only 40–80% in DED. Second, optimized intermediate post-processing, whether via in-process, post-hybrid, or eliminated heat treatments, must be rigorously studied to improve part microstructure and reduce machining forces. Third, the kinematic design of hybrid workstations, typically based on rigid multi-axis machining centers rather than less rigid robotic arms, defines process capability, and increased degrees of freedom boost part quality but complicate tool-path planning. Future work should explore specialty rotary tool paths, advanced slicing algorithms, and computer-aided manufacturing-integrated collision simulations to enable intelligent, collision-free hybrid machining. Finally, seamless hardware-software integration via cyber-physical systems, Internet of Things connectivity, and artificial intelligence-driven decision-making are critical for real-time monitoring and adaptive control of both DED and CNC operations. Embedding smart sensors within the HAM platform will facilitate continuous data exchange, autonomous problem-solving, and ultimately a truly intelligent manufacturing environment.

References

[1] Zhu Z, Dhokia VG, Nassehi A, Newman ST. A review of hybrid manufacturing processes – state of the art and future perspectives, Int J Comput Integr Manuf. 2013;26(7):596–615.

[2] Chu WS, Kim J, Lee YS, Choi Y, Kim G, Ahn SH. et al. Hybrid manufacturing in micro/nano scale: A review, Int J Precis Eng Manuf Technol. 2014;1(1):75–92.

[3] Kozak J., Rajurkar KP Hybrid machining process evaluation and development, In 2nd International Conference on Machining and Measurements of Sculptured Surfaces. 2000;501–536.

[4] Schuh G, Kreysa J, Orilski S. Roadmap "Hybride Produktion" Wie 1+1=3-Effekte in der Produktion Maximiert Werden Können, Z Wirtsch Fabrikbetr. 2009;104(5):385–391.

[5] Lauwers B, Klocke F, Klink A, Tekkaya AE, Neugebauer R, McIntosh D. Hybrid processes in manufacturing, CIRP Ann. 2014;63(2):561–583.

[6] Du W, Bai Q, Zhang B. A novel method for additive/subtractive hybrid manufacturing of metallic parts, Procedia Manuf. 2016;5:1018–1030.

[7] Kumar M, Sharma V. Assessment of microwave heat treatment on mechanical and microstructural behavior of Co-Cr alloy dental parts fabricated through selective laser melting, In All India Manufacturing Technology, Design and Research Conference. 2025;347–353.

[8] Kasperovich G, Hausmann J. Improvement of fatigue resistance and ductility of TiAl6V4 processed by selective laser melting, J Mater Process Technol. 2015;220:202–214.

[9] Song YA, Park S, Choi D, Jee H. 3D welding and milling: Part I – a direct approach for freeform fabrication of metallic prototypes, Int J Mach Tools Manuf. 2005;45(9):1057–1062.

[10] Rodriguez Franco C, Conje J. The evolution of the dialogue and perspectives on sustainable forest management with special emphasis on the United States of America, J Sustain For. 2023;42 (8):747–791.

[11] Kuhlman T, Farrington J. What is sustainability?, Sustainability. 2010;2(11):3436–3448.

[12] Halawa E. Sustainable energy: Concept and definition in the context of the energy transition – a critical review, Sustainability. 2024;16(4):1523.

[13] Giret A, Trentesaux D, Prabhu V. Sustainability in manufacturing operations scheduling: A state of the art review, J Manuf Syst. 2015;37:126–140.

[14] Pawanr S, Gupta K. A review on recent advances in the energy efficiency of machining processes for sustainability, Energies. 2024;17(15):3659.

[15] Madu CN, Kuei C, Madu IE. A hierarchic metric approach for integration of green issues in manufacturing: A paper recycling application, J Environ Manage. 2002;64(3):261–272.

[16] Prieto-Sandoval V, Jaca C, Ormazabal M. Towards a consensus on the circular economy, J Clean Prod. 2018;179:605–615.

[17] Herrmann C, Schmidt C, Kurle D, Blume S, Thiede S. Sustainability in manufacturing and factories of the future, Int J Precis Eng Manuf Technol. 2014;1(4):283–292.

[18] Choudhary N, Sharma V, Kumar P. Sustainability assessment framework of biomedical scaffolds: Additive manufacturing versus traditional manufacturing, J Clean Prod. 2023;418:138118.

[19] Badurdeen F, Jawahir IS. Strategies for value creation through sustainable manufacturing, Procedia Manuf. 2017;8:20–27.

[20] Kumar M, Sharma V. 3D printed bio-ceramic loaded PEGDA/vitreous Carbon composite: fabrication, characterization, and life cycle assessment, J Mech Behav Biomed Mater. 2023;143:105904.

[21] Reddy RDP, Elgazzar H., Sharma V Investigations of personalized and sustainable approach of oral drug delivery systems through additive manufacturing, Rapid Prototyp J. 2022;28(5):855–872.

[22] Singh AK, Sharma V. A comparative appraisal of sustainable strategy in ultrasonic assisted grinding of Nimonic 80A using novel green atomized cutting fluid, Sustain Mater Technol. 2022;32:e00423.

[23] Opoku RK. Championing green sustainability in manufacturing of a developing economy: Implications for sustainable performance through eco-regulatory compliance, Int J Product Perform Manag. 2025;74(7):2442–2470.

[24] Kreiger M, Pearce JM. Environmental life cycle analysis of distributed three-dimensional printing and conventional manufacturing of polymer products, ACS Sustain Chem Eng. 2013;1(12):1511–1519.

[25] Kokare S, Oliveira JP, Godina R. Life cycle assessment of additive manufacturing processes: A review, J Manuf Syst. 2023;68:536–559.

[26] Guo J, Zhao M, Wu X, Shi B, Santibanez Gonzalez EDR. Study on the distribution of PM emission rights in various provinces of china based on a new efficiency and equity two- objective DEA model, Ecol Econ. 2021;183:106956.
[27] Cao Y, Bian Y. Improving the ecological environmental performance to achieve carbon neutrality: the application of DPSIR-improved matter-element extension cloud model, J Environ Manage. 2021;293:112887.
[28] Liu W, Deng K, Wei H, Zhao P, Li J, Zhang Y. A decision-making model for comparing the energy demand of additive–subtractive hybrid manufacturing and conventional subtractive manufacturing based on life cycle method, J Clean Prod. 2021;311:127795.
[29] Watson JK, Taminger KMB. A decision-support model for selecting additive manufacturing versus subtractive manufacturing based on energy consumption, J Clean Prod. 2018;176:1316–1322.
[30] Priarone PC, Ingarao G, Lunetto V, Di Lorenzo R, Settineri L. The role of re-design for additive manufacturing on the process environmental performance, Procedia CIRP. 2018;69:124–129.
[31] Huang R, Riddle M, Graziano D, Warren J, Das S, Nimbalkar S. et al. Energy and emissions saving potential of additive manufacturing: the case of lightweight aircraft components, J Clean Prod. 2016;135:1559–1570.
[32] Liu Z, Islam F, Era IZ, Grandhi M. LCA-based environmental sustainability assessment of hybrid additive manufacturing of a turbine blade, Green Manuf Open. 2023;2:1.

Kapil Gupta

10 Process Capability Enhancement and Product Quality Improvement in Electrochemical Hybrid Machining for Various Materials

Abstract: Electrochemical hybrid machining (ECHM) processes are used for micro and precision machining, polishing, superfinishing, and machining glass-like ceramic materials. These processes are developed by hybridization of electrochemical machining (ECM) with other advanced machining processes and/or mechanical machining. Assistance from the magnetic field, vibration, and abrasives is also used to enable ECM to perform better. This chapter sheds light on the last 5 years selected important research on capability development, product quality enhancement, and optimization of ECHM processes. It has been analyzed that adopting various features in ECHM processes has resulted in surface quality improvement and process efficiency enhancement while processing a wide range of materials for aerospace, biomedical, and other industrial applications. Statistical, soft computing, and machine learning-based techniques have also been used to model and optimize ECHM with prolific outcomes. This chapter intends to facilitate researchers to gain knowledge and explore future avenues to establish the field further.

Keywords: ECDM, ECM, glass, hybrid machining, micromachining, optimization, surface quality

10.1 Introduction

Hybrid machining processes (HMPs) are the combination of two or more machining technologies to overcome their limitations and simultaneously utilize their benefits [1, 2]. These processes make use of both conventional and advanced machining operations where one is used for actual material removal and other is used as a medium to enhance its efficiency or productivity. Unique, crucial, and novel requirements related to the machine difficult material types, fabricate typical shapes and features, and perform operations to achieve superior surface quality are the major factors behind the evolution of HMPs. There are various types of HMPs and classified based on the machining principle involved. The major categories are combined and assisted types of HMPs. In combined category, electrochemical machining (ECM) is combined with laser, electric discharge, and mechanical machining such as grinding, milling, honing, and polishing. Electric discharge machining (EDM) is also combined with

 | https://doi.org/10.1515/9783111704456-010

laser and mechanical machining such as grinding, diamond grinding, and drilling. Whereas in assisted category, the assistance of vibration, heat, magnetic field, and abrasives is taken to enhance the capability of other advanced machining processes [3, 4].

ECM is a noncontact type of modern or advanced machining process in which material removal operation, for various purposes, takes place through anodic dissolution, after the generation of an electric field between a cathode tool and anode workpiece, under the influence of a suitable electrolyte [5, 6]. Faraday's laws of electrolysis govern the material removal in ECM. ECM's capabilities include machining of material of any hardness/toughness, production of high surface quality, generation of stress-free surfaces, and high productivity. In spite of those characteristics, ECM also suffers from limitations such as not suitable for electrically nonconducting materials, work surface damage due to chemical attack, performance depends on interelectrode gap and electrolyte characteristics. To overcome those limitations of ECM, electrochemical hybrid machining (ECHM) processes have been developed. The purpose behind development of ECHM processes is also to overcome the limitations related to the tool wear, surface quality, and productivity, of the advanced or mechanical machining, which is combined with ECM. Both combined and assisted type of ECHM processes are there. ECM combined with EDM results in electrochemical discharge machining (ECDM). ECM combined with mechanical machining operations results in electrochemical grinding (ECG), electrochemical honing (ECH), electrochemical polishing (ECPH), electrochemical milling. Assisted-type ECHM includes vibration-, abrasive-, laser-, and magnetic field-assisted ECM. All these ECHM processes, as shown in Fig. 10.1, have been developed to solve issues normally encountered when processing with a single modern machining and for special purposes such as precision finishing, machining of glass, micromachining, and polishing various mechanical parts and components made from aerospace alloys and other materials [7, 8].

Although these HMPs were developed a long time back, the last 5 years have seen significant breakthroughs in ECHM. This chapter presents the analysis of such attempts made by various researchers in the last 5 years to enhance the capability of ECHM, for productivity and product quality. It also briefly discusses the role of intelligent modeling and optimization techniques in that direction. There are few review studies available in this domain, either published a long back or specifically focused on one of the ECHM processes, mainly ECDM [9–13]. The novelty of this study lies in the fact that it covers some important types of ECHM processes and mainly emphasizes the last five years' development.

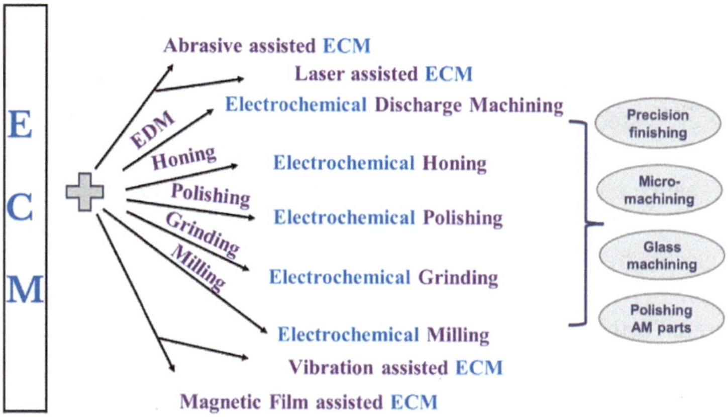

Fig. 10.1: Various important ECHM processes commonly used for aerospace alloys and other materials.

10.2 Review and Analysis of Recent Attempts

10.2.1 Process Capability Enhancement

Over the last 5 years, significant advancements have been achieved in enhancing the capability process of ECHM systems through strategic hybridization and integration of auxiliary mechanisms. These enhancements are primarily targeted at improving material removal rate (MRR), machining precision, efficiency, and operational stability when dealing with advanced engineering materials such as titanium alloys, stainless steel, nickel superalloys, and glass ceramics. ECHM combines the advantages of ECM with mechanical or thermal energy-assisted processes such as grinding, ultrasonic vibration, cryogenic flow, and abrasive flow. These hybrid integrations not only compensate for the limitations of conventional ECM (like low machining precision or poor flushing) but also significantly boost its process capability. Liu et al. [14] explored ECPH of Ti6Al4V alloy using high-speed microabrasive flow assistance, which substantially improved surface finish and enhanced localized material removal through mechanical abrasion superimposed on the electrochemical reaction. This hybridization not only accelerated polishing but also ensured better access to complex geometries, thereby enhancing the overall throughput. In the domain of ECDM, Wang et al. [15] developed a composite method integrating discharge milling and grinding for microstructure fabrication on glass materials. This dual mechanism allowed for better dimensional control and debris removal, resulting in more stable and accurate machining of fragile glass substrates.

Geng et al. [16] demonstrated another effective approach by introducing cryogenic airflow assistance in ECDM for drilling nickel-based superalloys. The use of cryogenically cooled air stream reduced tool wear, prevented thermal damage, and sustained the dielectric strength, which directly contributed to improving the precision and repeatability of the process. A multistage ECG technique for machining Ti6Al4V was also introduced for progressive material removal at optimized stages by overcoming the common issues of passivation and erratic spark behavior encountered in single-stage processes, facilitating enhanced depth control and superior surface integrity [17]. Another notable trend in recent research is the incorporation of vibration assistance, which has proven highly effective in improving flushing, avoiding electrode passivation, improving current density for smoother anodic dissolution, and stabilizing the electrochemical reactions [18, 19]. Improved localization of material removal and smoother machined profiles, especially in microhole and slot machining, were reported. Further, Che et al. [20] and Zhang et al. [21] extended the application of ultrasonic vibrations in polishing NiTi and copper materials respectively. These modifications enhanced electrolyte dynamics, improved ion exchange, and supported better feature replication in microfinishing applications. The advantages were most visible in complex miniature components where traditional ECM struggles due to ineffective flushing and non-uniform dissolution. In terms of abrasive-assisted enhancement, Mandal et al. [22] employed alumina abrasive powders in ECDM of maraging steel. This integration promoted mechanical erosion along with electrochemical reactions, ensuring more efficient material removal and reduced residual stress, which is often a challenge in conventional EDM or ECM.

These studies collectively highlight a clear shift towards hybrid process design, where multiple physical effects electrochemical, thermal, mechanical, and even cryogenic are integrated to expand the processing capabilities of ECM-based systems. This enables better machining of high-strength, temperature-resistant, and microfeatured components used in aerospace, biomedical, and automotive industries.

10.2.2 Product Quality Improvement

In recent years, there has been a remarkable focus on improving product quality through ECHM processes. ECHM, which merges ECM principles with various assisting techniques such as magnetic fields, ultrasonic vibrations, abrasives, and lasers, has shown great promise in enhancing surface integrity, dimensional accuracy, and surface roughness, particularly for advanced engineering materials and additively manufactured (AM) components. Electrochemical processes inherently offer burr-free, stress-free, and non-thermal machining. However, their effectiveness in achieving ultra-fine surface finishes and precise microfeatures is often limited by issues such as poor flushing, passivation, and low localization. To address these challenges, researchers have explored hybrid strategies that not only improve the material removal mechanism but also significantly upgrade the final product's quality in terms of surface smoothness, structural uniformity, and functional performance.

One major application area where ECHM has demonstrated substantial improvement is electrochemical deburring. He et al. [23] applied ECHM to effectively deburr microdrilled holes on 304 stainless steel plates. This process enabled smooth and consistent removal of burrs from hard-to-access internal edges, which is essential for the functional performance of precision components. Similarly, a moving cathode-based electrochemical deburring method studied by He et al. [24] offered precise and localized burr removal from stainless steel sheets, enhancing the geometric accuracy and edge definition of microfeatures. The integration of magnetic field assistance in ECM is yet another significant advancement. Wang et al. [25] used magnetic field-assisted ECM on 2Cr13 stainless steel and achieved notable improvements in surface morphology. The technique reduced average surface roughness to as low as 0.28 µm, showcasing its potential for high-quality finishing in industrial components. In a similar study, Wang et al. [26] applied magnetic field-assisted ECPH to TC4 titanium alloy and achieved more than a 10-fold reduction in surface roughness, along with a 40% improvement in polishing efficiency. These results emphasize the benefits of magnetic agitation in enhancing electrolyte flow and achieving more uniform material dissolution.

The trend of applying ECHM to AM parts has gained strong momentum. AM techniques such as laser powder bed fusion (LPBF) often result in rough and irregular surfaces that require post-processing [28]. Liu et al. [29] and Zhao et al. [31] demonstrated that ECPH of LPBF-fabricated AlMgScZr and 316L stainless steel components could achieve up to 90% improvement in mean roughness depth, providing smooth surfaces suitable for high-precision and functional applications. Moreover, Liu et al. [30] showed that laser-electrochemical hybrid polishing not only improved surface finish but also enhanced wear resistance, which is vital for components used in biomedical and structural applications. Additionally, Chenhao Zhao et al. [32] introduced electrochemical mechanical polishing to smoothen internal holes in 304 stainless steel developed by selective laser melting (SLM). This hybrid method effectively reduced areal average roughness and eliminated surface inconsistencies within complex geometries. It illustrates the feasibility of combining ECM with mechanical abrasion for finishing hard-to-reach internal surfaces. Whether it is for improving biocompatibility, wear resistance, aesthetic appearance, or dimensional accuracy, ECHM offers a versatile toolbox to address post-machining challenges in modern manufacturing. Table 10.1 summarizes the recent attempts carried out towards process capability and product quality improvement in ECHM process domains.

10.2.3 Effects of Modeling and Optimization Techniques

Optimization of manufacturing process is one of the techniques to enhance their performance to obtain improved part quality and process productivity. Machine learning-based optimization has got special attention in the last 3 years for productivity and quality enhancement in manufacturing processes, developments in performance enhancement of electrochemical-based HMPs by modeling and optimizing. Implementation and effect of both statistical and soft computing-based techniques on the performance of ECHM processes have been highlighted.

Tab. 10.1: Summary of developments in ECHM for process capability enhancement and product quality improvement.

	Process capability enhancement	
References	**Process, material, product**	**Features related to capability enhancement**
[14]	Electrochemical polishing, Ti6Al4V alloy	High-speed flow of microabrasives
[15]	Electrochemical discharge machining, glass	Combining electrochemical discharge milling and grinding
[16]	Electrochemical discharge drilling, Ni superalloy	Cryogenic air flow assistance
[17]	Electrochemical grinding, Ti6Al4V	Multistage operation
[18]	Electrochemical machining, 301 stainless steel	Vibration assistance
[19]	Wire electrochemical micromachining, NiTi alloy	Vibration assistance
[20]	Electrochemical polishing, NiTi	Vibration assistance
[21]	Electrochemical polishing, Cu	Vibration assistance
[22]	Electrochemical discharge machining, maraging steel	Abrasive assistance
	Product quality improvement	
Reference	**Process, material**	**Findings**
[23]	Electrochemical deburring, 304 stainless steel plate	Effective burr removal from the edges of the drilled hole
[24]	Moving cathode based electrochemical deburring, 304 stainless-steel cheet	Precise and effective removal of burr material
[25]	Magnetic field assisted electrochemical machining, 2Cr13 stainless steel	Surface morphology improvement with 0.28 µm average surface roughness
[26]	Magnetic field-assisted electrochemical polishing, TC4 titanium alloy	More than 10 times reduction in surface roughness Increased efficiency by 40%
[27]	Tool-vibration-assisted pulsed electrochemical machining, γ-TiAl TNM	Good machining accuracy and surface quality of aero engine blades made of γ-TiAl TNM alloy processed at optimum parameters

Tab. 10.1 (continued)

Process capability enhancement		
References	**Process, material, product**	**Features related to capability enhancement**
[29]	Electrochemical polishing, LPBF fabricated AlMgScZr alloys	Improved surface quality
[30]	Laser electrochemical hybrid polishing, SLM fabricated 316L stainless steel	Improved surface finish and wear resistance
[31]	Solid dielectric electrochemical polishing LPBF fabricated 316L stainless steel	More than 90% improvement in mean roughness depth
[32]	Electrochemical mechanical polishing, SLM fabricated 304 stainless steel	Significant reduction in areal average roughness

An identification of optimal parameters was successfully accomplished in research on ECDM of borosilicate glass for hole fabrication [33]. Gray relational-grade analysis provided as an optimal combination of process parameters 36 V, 110 µm/min tool feed rate, and 1-minute machining time with nickel-coated tool and 2 minutes with bare copper tool, for the best hole quality. It was also found that voltage and machining time affected the hole geometry greatly. Technique for Order Preference by Similarity to Ideal Solution (TOPSIS) was used to optimize the parameters of ultrasonic-assisted rotary micro-ECDM for minimizing circularity error and overcut of borosilicate glass substrate [34]. Vibration amplitude 6 µm, rotational speed 80 rpm, and feed rate 0.7 mm/min produced simultaneous least values of overcut – 153 µm and best value of circularity – 0.889. Another important study presents the multiperformance optimization of micro-ECDM for producing quality microchannels on borosilicate glass [35]. Taguchi-based optimization helped to obtain optimal set of process parameters as 50 V voltage, 25 wt% electrolyte concentration, 0.5 duty ratio and 50 Hz pulse frequency for the best values of MRR, surface roughness, and width of cut. Performance of micromachining of silica glass using ECDM was improved using particle swarm optimization [36]. Silica glass with low heat-affected zone and wear of cutting tool was successfully machined. ECDM of borosilicate glass with moving wire was optimized using Taguchi technique [37]. The combination of 35 V, 25% electrolytic concentration, and 20 mm inter-electrode gap resulted in fabrication slot with the maximum aspect ratio of ~32 at stable and wire-breakage-free machining. Principal component-based gray relational analysis was implemented to obtain a better performance of ECG of Ti grade 5 alloy [38]. Fifty-four experiments were conducted, and correspondingly, surface roughness and MRR were measured. Parameter optimization resulted in the best values of conflicting responses, i.e., MRR – 0.0395 g/min and surface roughness – 2.02 µm.

10.3 Conclusion

With a continued focus on process control, intelligence, and sustainability, ECHM has emerged as a key enabling technology in advanced manufacturing sectors serving the aerospace, biomedical, and other industries. After studying and analyzing the recent developments in electrochemical-based HMPs, the following conclusions can be drawn:

- The process capability enhancement in ECHM has been driven by intelligent hybridization strategies, innovative auxiliary techniques, and data-driven optimization frameworks. These advances not only improve productivity but also ensure sustainable and precise machining for next-generation materials and components.
- Product quality improvement in ECHM is being achieved through a range of hybrid techniques that enhance electrochemical reactions, increase process control, and enable high-precision finishing. These advancements have broadened the scope of ECHM from traditional micromachining to complex surface treatment applications, especially in aerospace, biomedical, and AM-based industries.
- The use of advanced modeling and optimization tools (e.g., Taguchi methods, gray relational analysis, and machine learning algorithms) has enabled researchers to finely tune the hybrid processes for maximal performance. The increased process stability, dimensional accuracy, and material compatibility derived from ECHM prove its recognition as a versatile manufacturing tool.
- Some promising future research avenues include exploring hybrid optimization to enhance the outputs of ECHM processes, extending machine learning-based modeling and optimization to other HMPs, investigating the impact of optimization techniques on sustainability indicators such as energy and resource consumption and emissions.
- Future research can also be directed to investigate fabrication and fatigue characteristics of parts, such as landing gears and airframes, and precision machining of jet engine parts, especially made of titanium alloys, by ECHM processes.

References

[1] Singh R, Davim JP. Non-conventional hybrid machining processes: theory and practice. CRC Press; 2021.

[2] Lauwers B, Klocke F, Klink A, Tekkaya AE, Neugebauer R, McIntosh D. Hybrid processes in manufacturing, CIRP Ann. 2014;63(2):561–583.

[3] Grzesik W, Ruszaj A. Hybrid manufacturing processes – physical fundamentals, modelling and rational applications. Springer; 2021.

[4] Singh G, Singh R, Rao PS. On modes of electrochemical machining: a comprehensive review, Adv Mater Process Technol. 2024;1–39. https://doi.org/10.1080/2374068X.2024.2439703

[5] Liu S, Thangamani G, Thangaraj M, et al. Recent trends on electrochemical machining process of metallic materials: A review, Arch Civ Mech Eng. 2023;23:158. https://doi.org/10.1007/s43452-023-00703-w
[6] Pârvu GM, Enciu C, Ghiculescu LD. A review on electrochemical machining and micromachining, Acta Tech Napoc. 2021;64(4s):745–754.
[7] Painuly M, Singh RP, Trehan R. Electrochemical machining and allied processes: a comprehensive review, J Solid State Electrochem. 2023;27:3189–3256. https://doi.org/10.1007/s10008-023-05610-x
[8] Pawar AB, Chavan SS, Jadhav PV, Jadhav DB. Emerging trends in hybrid ECM: A review of multi-assisted machining techniques, Eng Res Express. 2025;7(1):012504. https://doi.org/10.1088/2631-8695/adb53e
[9] Oza AD, Goyal A, Buch V, Kumar M. Electrochemical discharge machining process: A review on process parameters and future scope, Mater Today Proc. 2022;62(13):6956–6961. https://doi.org/10.1016/j.matpr.2021.12.341
[10] Prakash A, Kumar A, Ballav R. Process capability of electrochemical discharge machining: A review, Proc Inst Mech Eng L. 2023;238(2):282–300. https://doi.org/10.1177/14644207231218386
[11] Yadav RN. Electrochemical spark machining–based hybrid machining processes: Research trends and opportunities, Proc Inst Mech Eng B. 2018;233(4):1037–1061. https://doi.org/10.1177/0954405418755825
[12] Grover S, Singh S, Mangal SK. Hybridization of μ-electrochemical discharge machining (μ-ecdm) process: A review, Mater Today Proc. 2023;80(2):499–508. https://doi.org/10.1016/j.matpr.2022.10.204
[13] Srivastava A, Yadav SKS. Machining issues on electrochemical spark machining a Review, Mater Today Proc. 2020;26(2):2853–2861. https://doi.org/10.1016/j.matpr.2020.02.593
[14] Liu J, Wang Z, Xu Z. Electrochemical polishing of Ti6Al4V alloy assisted by high-speed flow of micro-abrasive particles in NaNO3 electrolyte, Materials (Basel). 2022;15(22):8148. https://doi.org/10.3390/ma15228148
[15] Wang T, Liu Y, Wang K, Lv Z. Investigation on a sustainable composite method of glass microstructures fabrication Electrochemical Discharge Milling and Grinding (ECDM-G), J Clean Prod. 2023;387:135788. https://doi.org/10.1016/j.jclepro.2022.135788
[16] Geng T, Xu Z, Lu J. A sustainable, high-quality method for machining film cooling holes: cryogenic airflow-assisted electrochemical discharge drilling, J Clean Prod. 2023;426:139017. https://doi.org/10.1016/j.jclepro.2023.139017
[17] Yang G, Ming P, Niu S, Qin G, Liu H, Li D, Zhang A. An investigation of the efficient precise continuous electrochemical grinding process of Ti–6Al–4V, Materials (Basel). 2024;17(8):1729. https://doi.org/10.3390/ma17081729
[18] Shen ZY, Tsui HP. An investigation of ultrasonic-assisted electrochemical machining of micro-hole array, Processes. 2021;9(9):1615. https://doi.org/10.3390/pr9091615
[19] Hung JC, Yang PJ. Electrochemical microslot machining by ultrasonic-vibration-aided electrolyte on nitinol wire, Processes. 2021;9(10):1752. https://doi.org/10.3390/pr9101752
[20] Che J, Shi G, Xu Y. Experimental study on ultrasonic-assisted electrochemical polishing of NiTi alloy, Phys Scr. 2024;99(4):045973.
[21] Zhang X, Wang J, Chen J, Lyu B, Yuan J. Material removal and surface modification of copper under ultrasonic-assisted electrochemical polishing, Processes. 2024;12(6):1046. https://doi.org/10.3390/pr12061046
[22] Mandal N, Kumar N, Das AK. Machining of maraging steel through Al2O3 abrasive powder-assisted electrochemical discharge machining, In Shunmugam MS, Doloi B, Ramesh R, and Prasanth AS editors. Advances in Modern Machining Processes. Lecture Notes in Mechanical Engineering. Singapore: Springer; 2023. https://doi.org/10.1007/978-981-19-7150-1_9

[23] He T, Huang S, Fang S, Deng H, Ding Q, Zhang M. Study on the electrochemical deburring for the external surface of the microhole caused by mechanical drilling process, Machines. 2022;10(9):726. https://doi.org/10.3390/machines10090726
[24] He T, Li C, Ding Q, et al. Research on electrochemical micropore deburring based on moving cathode, J Mech Sci Technol. 2025;39:3359–3369. https://doi.org/10.1007/s12206-025-0529-2
[25] Wang M, Xu J, Fan D, et al. The improvement of the surface morphology via magnetic field-assisted electrochemical machining, Int J Adv Manuf Technol. 2024;133:1823–1831. https://doi.org/10.1007/s00170-024-13944-7
[26] Wang Y, Liu H, Yin X, Yi Y, Feng M. Study on magnetic field–electrochemical composite polishing mechanism of TC4 titanium alloy, Phys Scr. 2023;100(7):075002. https://doi.org/10.1088/1402-4896/addca3
[27] Liu J, Liu Y, Zhang Z, Wang H. Parameter optimization and experimental study on tool-vibration-assisted pulsed electrochemical machining of γ-TiAl TNM blades, Appl Sci. 2022;12(16):8042.
[28] Mu J, Sun T, Leung CLA, Oliveira JP, Wu Y, Wang H, Wang H. Application of electrochemical polishing in surface treatment of additively manufactured structures: A review, Prog Mater Sci. 2023;136:101109. https://doi.org/10.1016/j.pmatsci.2023.101109
[29] Li Y, Huang G, Hu W, Yang T, Zeng Y. Optimizing internal surface quality of laser powder bed fusion AlMgScZr alloy parts by electrochemical polishing techniques, Mater Today Commun. 2025;42:111513. https://doi.org/10.1016/j.mtcomm.2025.111513
[30] Liu J, Li C, Yang H, Liu J, Wang J, Deng L, Fang L, Yang C. Study on laser–electrochemical hybrid polishing of selective laser melted 316L stainless steel, Micromachines. 2024;15(3):374. https://doi.org/10.3390/mi15030374
[31] Liu S, Li C, Jin X, Jiang W, Cao X, Liu G, Guo Z, Yang YX, Ong LCL. Quantitative-regulated Material Removal Rate in Solid Dielectric Electrochemical Polishing (QRR-SDEP) for smoothing high roughness surface of additively manufactured 316L stainless steel components, Addit Manuf. 2023;73:103689. https://doi.org/10.1016/j.addma.2023.103689
[32] Zhao C, Qu N, Tang X. Electrochemical mechanical polishing of internal holes created by selective laser melting, J Manuf Process. 2021;64:1544–1562. https://doi.org/10.1016/j.jmapro.2021.03.003
[33] Tiwari AK, Panda SS. Optimization of process parameters in ECDM machining using taguchi based grey relation analysis, Measurement. 2023;216:112971. https://doi.org/10.1016/j.measurement.2023.112971
[34] Grover S, Mangal SK, Singh S Micro-machining and process optimization of ultrasonic assisted rotary µ-electrochemical discharge machining using TOPSIS method. Mater Today Proc. In Press. https://doi.org/10.1016/j.matpr.2023.01.001
[35] Mondal K, Mallick B, Sarkar MK, Biswas SK, Mahato JK. Optimization of micro-ECDM performances during micro-channeling on glass using taguchi technique, Mater Today Proc. 2024. https://doi.org/10.1016/j.matpr.2024.05.014
[36] Mondal K, Hikmat SK, Mallick B, Mahato J, Dutta P. Evolutionary algorithms-based optimization of electrochemical discharge micromachining process during micro-channel fabrication on glass, Mater Today Proc. 2023;80(2):613–619. https://doi.org/10.1016/j.matpr.2022.11.058
[37] Kumar M, Vaishya RO, Suri NM, Gupta A. Parametric optimization of traveling wire electrochemical discharge machining (TW-ECDM) process for aspect ratio during machining of borosilicate glass, Mater Today Proc. 2022;64(3). https://doi.org/10.1016/j.matpr.2022.03.599
[38] Yadav SK, Yadav SKS. Multi-objective optimization of electrochemical cut-off grinding process of Ti-6Al-4V using PCA-based grey relational analysis, Mater Today Proc. 2020;22:3089–3099.

Index

 | https://doi.org/10.1515/9783111704456-011